Springer Proceedings in Materials

Volume 23

Springer Proceedings in Materials publishes the latest research in Materials Science and Engineering presented at high standard academic conferences and scientific meetings. It provides a platform for researchers, professionals and students to present their scientific findings and stay up-to-date with the development in Materials Science and Engineering. The scope is multidisciplinary and ranges from fundamental to applied research, including, but not limited to:

- Structural Materials
- Metallic Materials
- Magnetic, Optical and Electronic Materials
- Ceramics, Glass, Composites, Natural Materials
- Biomaterials
- Nanotechnology
- Characterization and Evaluation of Materials
- Energy Materials
- Materials Processing

To submit a proposal or request further information, please contact one of our Springer Publishing Editors according to your affiliation:

European countries: **Mayra Castro** (mayra.castro@springer.com)
India, South Asia and Middle East: **Priya Vyas** (priya.vyas@springer.com)
South Korea: **Smith Chae** (smith.chae@springer.com)
Southeast Asia, Australia and New Zealand: **Ramesh Nath Premnath** (ramesh.premnath@springer.com)
The Americas: **Michael Luby** (michael.luby@springer.com)
China and all the other countries or regions: **Mengchu Huang** (mengchu.huang@springer.com)
This book series is indexed in **SCOPUS** database.

Ramzi Khiari · Mohammad Jawaid
Editors

Proceedings of the 3rd International Congress of Applied Chemistry & Environment (ICACE-3)

Advances in Chemistry and Clean Processes on Materials and Nanomaterials

Springer

Editors
Ramzi Khiari
Higher Institute of Technological Studies (ISET) of Ksar-Hellal, Department of textile
ISET Ksar Hellal, University of Monastir
Monastir, Tunisia

LGP2, Grenoble INP, CNRS
University of Grenoble Alpes
Grenoble, France

Mohammad Jawaid
Laboratory of Biocomposite Technology
Universiti Putra Malaysia
Serdang, Selangor, Malaysia

ISSN 2662-3161 ISSN 2662-317X (electronic)
Springer Proceedings in Materials
ISBN 978-981-99-1967-3 ISBN 978-981-99-1968-0 (eBook)
https://doi.org/10.1007/978-981-99-1968-0

This Springer imprint is published by the registered company Springer Nature Singapore Pte Ltd.
The registered company address is: 152 Beach Road, #21-01/04 Gateway East, Singapore 189721, Singapore

Preface

Dear Colleagues,

The third International Congress of Applied Chemistry and Environment (ICACE–3) organized by the Laboratory of Environmental Chemistry and Clean Process (LCE2P—LR21ES04—Faculty of Sciences of Monastir—Tunisia), in partnership with the ATuTeD (Tunisian Association of Sustainable Technologies) and the SFGP (French Society of Process Engineering) will take place from May 27th to 29th 2022. This event is an opportunity for participants to present their research and will provide a framework for fruitful scientific exchanges between academics and industrialists. Paper proposals may fall under one of the following thematic sections:

- Environment and Climate Change,
- Green Chemistry and Natural Substances,
- Biobased Materials and Nanomaterials,
- Sustainable Textiles and Clean Processes.

This proceeding comprises of illustrations and tables summarize up to date information on research carried out production of engineering at different fields. This proposal proceeding collet the information and knowledge of new way to intends to promote sharing ideas and emerging technologies, as well as to foster research and development collaborations amongst academia, research institutions and relevant industries. The main objective of this international event is to promote innovation and creativity among researchers working in the engineering and Materials research and development.

Monastir, Tunisia — Dr. Ramzi Khiari
Serdang, Malaysia — Prof. Mohammad Jawaid

Contents

About the Editors

Dr. Ramzi Khiari is a senior lecturer at the Higher Institute of Technological Studies in Ksar-Hellal (Monastir, Tunisia) in the department of Textile Engineering. He was graduated in 2005 at the National Engineering School of Monastir in the specialty of Textile Chemistry, before getting a Master degree (2007) from the same institution. Then, he performed a sandwich Ph.D. thesis (2010) between ENIM from University of Monastir in Tunisia and Grenoble INP in France. Finally, in 2017, he has got the diploma of "Habilitation Universitaire" from University of Monastir and 2020 he has got the diploma "Habilitation à diriger des Recherches" in Grenoble INP. His research interests focus on the valorization of biomass at multi-scale levels namely: fibers, nanocellulose, lignin, hemicelluloses and their used as potential raw material in several industrial applications (Textile, papermaking, polymeric materials, composites and nanocomposites). A particular focus is given for vegetal biomass from annual plants, and particularly agricultural residues and industrial wastes. During his career, he coordinated several research projects, mainly projects with industry, and supervised various post-graduation students. He has been member of scientific committees, organizing committees and participated in national/international evaluation boards.

Dr. Mohammad Jawaid is currently working as Senior Fellow (Professor) at Biocomposite Technology Laboratory, Institute of Tropical Forestry and Forest Products (INTROP), Universiti Putra Malaysia (UPM), Serdang, Selangor, Malaysia, and also has been Visiting Professor at the Department of Chemical Engineering, College of Engineering, King Saud University, Riyadh, Saudi Arabia since June 2013. Recently he joined as Distinguish Visiting Professor at Malaysia Japan International Institute of Technology (MJIIT)-Universiti Teknologi Malaysia (UTM), Kuala Lumpur, Malaysia. He has more than 22 years of experience in teaching, research, and industries. His area of research interests includes hybrid composites, lignocellulosic reinforced/filled polymer composites, advance materials: graphene/nanoclay/fire retardant, modification and treatment of lignocellulosic fibers and solid wood, biopolymers and biopolymers for packaging applications, nanocomposites and nanocellulose fibers, and polymer blends. So far, he has published 60 books, 80 book chapters, more than 450 peer-reviewed international journal papers, and several published review papers under top 25 hot articles in science direct during 2013–2020. He also obtained 6 Patents and 6 Copyrights/Industrial design. H-index and citation in Scopus are 75 and 25736 and in Google scholar, H-index and citation are 89 and 34711; H-index = 70 (Web of Science). He is founding Series Editor of Composite Science and Technology Book Series, Sustainable Materials and Technology book series, and Smart Nanotechnology book series from Springer-Nature, and also Series Editor of Springer Proceedings in Materials, Springer-Nature. Beside that he is also member of International Advisory Board of Springer Series on Polymer and Composite Materials. He worked as guest editor of special issues of SN Applied Science, Current Organic Synthesis and Current Analytical Chemistry, International Journal of Polymer Science, IOP Conference Proceeding. He also in Editorial Board Member of Journal of Polymers and The Environment, Journal of Plastics Technology, Applied Science and Engineering Progress Journal, Journal of Asian Science, Technology and Innovation and the Recent Innovations in Chemical Engineering. He is life member of Asian Polymer Association, and Malaysian Society for Engineering and Technology.

He is also member of Society of Plastics Engineers, American Chemical Society, Athens Institute for Education and Research, and International Association of Advanced Materials. Besides that, he is also reviewer of several high-impact international peer-reviewed journals of Elsevier, Springer, Wiley, Saga, ACS, RSC, Frontiers, etc. Presently, he is supervising 7 Ph.D. students (4 Ph.D. as Chairman, and 3 Ph.D. as Member) and 3 Master's as Member in the fields of hybrid composites, green composites, nanocomposites, natural fiber-reinforced composites, nanocellulose, etc. 32 Ph.D. and 14 Master's students graduated under his supervision in 2014–2022. He has several research grants at university, national, and international levels on polymer composites of around 6 million Malaysian ringgits (USD 1.36 Million). He also delivered plenary and invited talks in international conferences related to composites in India, Turkey, Malaysia, Thailand, the United Kingdom, France, Saudi Arabia, Egypt, and China. Besides that, he is also a member of technical committees of several national and international conferences on composites and material science. Recently Dr. Mohammad Jawaid received Excellent Academic Award in Category of International Grant-Universiti Putra Malaysia-2018 and also Excellent Academic Staff Award in industry High Impact Network (ICAN 2019) Award. Beside that Gold Medal-Community and Industry Network (JINM Showcase) at Universiti Putra Malaysia. He also won Silver medals in Malaysian Technology expo (MTE-2021), MTE-SDG International Innovation Awards-2021 and International Technology Expo (ITEX-2022). He also Received Publons Peer Review Awards 2017, and 2018 (Materials Science), Certified Sentinel of science Award Receipient-2016 (Materials Science) and 2019 (Materials Science and Cross field). He is also Winner of Newton-Ungku Omar Coordination Fund: UK-Malaysia Research and Innovation Bridges Competition 2015.

Antibacterial Activity of Polyamide and High-Density Polyethylene Nets Grafted with Polyethylene Glycol

Intissar Amara, Wafa Miled, Rihab Ben Slama, Pascale Chevallier, Diego Mantovani, Claire Hellio, and Néji Ladhari

Abstract The aims of this study are to stop biofouling formation and to create antibacterial layers on aquaculture nets using a chemical product known for its antibacterial behavior. Polyamide 6.6 and high-density polyethylene nets were grafted with polyethylene glycol by following two methods. At first, grafted nets were evaluated with SEM and XPS in order to study the morphological and chemical information. Then, the antibacterial activity was studied at laboratory scale by measuring the rate of adhesion, on nets, of three bacterial strains (Pseudoalteromonas citrea, Pseudoalteromonas elyakovii, and Vibrio harveyi). Results highlighted that the surface of the ungrafted nets was the most colonized by bacteria, with a progressive increase in bacterial adhesion over the time. For PA 6.6 and HDPE nets, the best antibacterial behavior was noted on nets grafted with PEG after surface activation at 45 °C.

Keywords Aquaculture nets · Biofouling · Aquatic bacteria · Grafting

I. Amara (✉) · W. Miled
Textile Engineering Laboratory, University of Monastir, Monastir, Tunisia
e-mail: intissar.amara@hotmail.fr

W. Miled
Department of Early Childhood, University College Taraba, Taif University, Taif, Saudi Arabia

R. B. Slama
Laboratory of Analysis, Treatment and Valorization of Pollutants of the Environment and Products, Faculty of Pharmacy, University of Monastir, Monastir, Tunisia

P. Chevallier · D. Mantovani
Laboratory for Biomaterials and Bioengineering, University Laval, Quebec, QC G1V 0A6, Canada

C. Hellio
Université de Brest, CNRS, IRD, Ifremer, LEMAR, Institut Universitaire Européen de la Mer, 29280 Plouzané, France

N. Ladhari
Higher Institute of the Fashion Trades of Monastir, University of Monastir, Monastir, Tunisia

University of Business and Technology, Jeddah, Saudi Arabia

R. Khiari and M. Jawaid (eds.), *Proceedings of the 3rd International Congress of Applied Chemistry & Environment (ICACE-3)*, Springer Proceedings in Materials 23, https://doi.org/10.1007/978-981-99-1968-0_1

1 Introduction

The marine biofouling is a natural phenomenon which consists on the colonization of all immersed surfaces, in marine environment, by aquatic organisms [1, 17, 24, 25]. The process of biofouling, which scenario is modeled as shown in Fig. 1, takes place generally in four main stages: primary film formation, biofilm formation, eukaryote settlement, and macro-organisms' development [8, 10, 13, 17, 24]. Immediately after the immersion of a surface in seawater, the primary film formation starts, within minutes, and it is due to the adsorption of proteins, organic, and inorganic macromolecules which exist in seawater [17, 24, 25].

Such a natural phenomenon affects all man-made immersed structures. Consequently, aquaculture field is widely affected [16]. In fact, this spontaneous phenomenon causes the block of net openings and this blockage reduces oxygen supply, nutrient exchange, and removal of waste products [11, 19]. As a consequence, the health of farmed species, such as mussels, oysters, and fishes will be affected. Several studies showed that biofouling can reduce farmed shellfish growth, weight, and survival. For example, the fish Ectopleura larynx and the anemone Anthothoe albocincta are severely affected by cnidarian biofouling and their gill and skin get damaged [4–6]. To reduce biofouling problems, a lot of antifouling biocides have been developed and applied as antifouling paints for boats and for aquaculture nets [12, 26]. Unfortunately, they were found to be harmful not only to bacterial development, but also to different aquatic species and they were so toxic to the environment [1].

In this study, polyamide 6.6 and high-density polyethylene nets were grafted with polyethylene glycol (PEG) by following two methods in order to modify the surface properties. In fact, PEG was chosen for its ability to form a hydrated layer to resist protein adhesion [18]. And since protein adhesion presents the first stage in biofouling formation, the reduction of protein adsorption causes the reduction of other micro- and macro-organisms' adhesion. The antibacterial efficiency of grafted

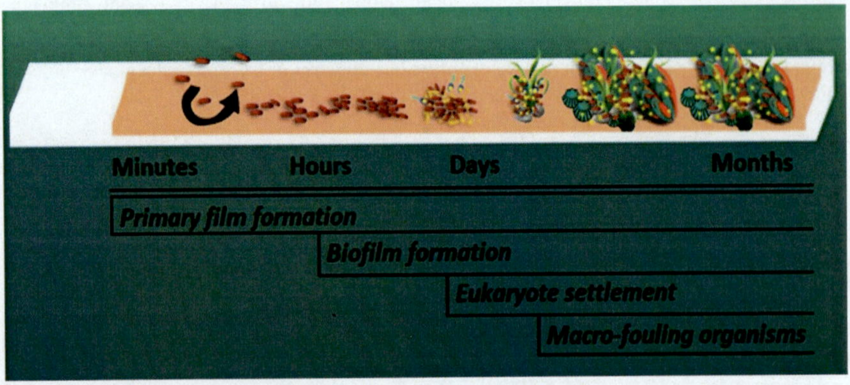

Fig. 1 Schematic representation of fouling formation sequences

nets was studied using a new original method. This is the first time the epifluorescence microscopy "Zeiss Vivatome", and the software "Image J" were used to study the adhesion of aquatic bacteria on aquaculture nets. So, after immersion of samples in three different bacterial strains (Pseudoalteromonas citrea, Pseudoalteromonas elyakovii, and Vibrio harveyi), they were observed with an epifluorescence microscopy (Zeiss Vivatome) in order to detect adhered bacteria. The resulting images were then processed by image analysis software called "Image J". PA 6.6 and HDPE nets were evaluated using scanning electron microscope (SEM) and X-ray photoelectron spectrometer (XPS).

2 Experimental

2.1 Materials

Samples of polyamide 6.6 (PA 6.6) and high-density polyethylene (HDPE) nets were purchased from the fish farm "RAFAHA" (Monastir, Tunisia) and used as the basic material. Polyamide 6.6 and high-density polyethylene nets have knotless hexagonal meshes with a length of the mesh's side equal to 1.5 cm. Besides, an average diameter of filaments equal to 27 μm and 18 μm, respectively for PA 6.6 and HDPE samples. Those nets had been firstly cleaned with a standard scouring step, in order to remove all impurities that may exist on the net samples.

2.2 Chemicals

For Polyethylene glycol bis(carboxymethyl) ether (PEG) ($HOOCCH_2(OCH_2CH_2)_nOCH_2COOH$) grafting process, many chemical products had been used and they are dopamine ($C_8H_{11}NO_2$, M = 189.64 g/mol), EDAC (1-[3-diméthylaminopropyl]-3-éthylcarbodiimide chlorhydrate: $C_8H_{17}N_3$, M = 191.70 g/mol), Tampon MES ($C_6H_{13}NO_4S$, M = 195.24 g/mol), and Tampon Tris ($C_4H_{11}NO_3$, M = 121.14 g/mol). All these products were supplied from Sigma-Aldrich.

2.3 PEG Grafting Methods

Polyethylene glycol is a hydrophilic material that can create interfacial water layers on it. Those layers prevent any direct contact between the surface and the protein, so this property can be valued to reduce bacterial adhesion.

Before grafting, the surfaces of the PA 6.6 and HDPE were functionalized using dopamine, which acts as an anchor point to ensure the grafting of other products. First of all, samples were immersed in dopamine solution (0.2 mol/L). To carry out this reaction, two temperatures were studied: room temperature (22 °C) and 45 °C. The immersion time was for 18 h both. The dopamine solution was prepared by dissolving dopamine in Tris buffer solution (10 mM) with a pH adjusted to 8.5 using HCl (initial pH was equal to 8.6). Samples were then rinsed several times with nanopure water and finally dried.

After that, PA 6.6 and HDPE net samples were treated in 0.1 g/mL of PEG solution (pH 4.75, MES buffer), previously activated with EDAC (3 mg/mL) for 3 times (10 min each). After 1 h of reaction under stirring, nets were rinsed three times with MES buffer and five times with deionized water. Finally, they were dried and stored under vacuum [22].

In the rest of this manuscript, the terms "PA-Dop-PEG", "PA-Dop45°-PEG", "HDPE-Dop-PEG", and "HDPE-Dop45°-PEG" refer, respectively, to PA 6.6 and HDPE nets grafted with PEG after dopamine activation at room temperature (22 °C) and at 45 °C.

3 Surface Characterization

3.1 Scanning Electron Microscope Analyses

A scanning electron microscope SEM (FEI-Quanta 250, Oregon, USA) operating in low-vacuum mode [2] was used in order to analyze surface morphologies of ungrafted and grafted nets. This microscope defines qualitatively the distribution of the product on the surface. Nets were cut into small pieces (8 mm), and a thin layer of gold was spread on those pieces in order to make them conductive. After that, nets were analyzed with SEM and three different positions were observed with 1000 times magnification at a voltage of 15 kV, for each grafted net.

3.2 X-ray Photoelectron Spectrometer Analyses

An X-ray photoelectron spectrometer XPS (PHI5600-ci spectrometer, Physical Electronics USA, MN, USA) was employed to determine chemical composition on the surface. Three samples of each net were analyzed using XPS. Then, the percentage of each chemical compound was determined and finally, the average was calculated.

The theoretical compositions usually found with XPS are

- For pure PA 6.6 ($(C_{12}H_{22}N_2O_2)_n$): 75.0% of carbon, 12.5% of oxygen and 12.5% of nitrogen.
- For pure PEHD ($(C_2H_4)_n$): 100% of carbon.

- For PEG (C_2H_4O): 66.66% of carbon and 33.33% of oxygen.

XPS low resolution results will be compared with those compositions in order to evaluate the surface modifications. Also, quantitative analysis was carried out using the C 1s peak area on control and grafted samples.

4 Antibacterial Activity Evaluation

4.1 Preparation of Strains of Bacteria

Antibacterial behavior of grafted nets was studied at laboratory scale. Therefore, three strains of bacteria were used and they are *Pseudoalteromonas citrea* (ATCC 29270), *Pseudoalteromonas elyakovii* (ATCC 700519), and *Vibrio harveyi* (ATCC 35084) [7].

Those strains are generally used as model species for marine biofouling studies.

To prepare bacteria strains, selected bacteria were pre-grown, at 20 °C, in sterile Marine Bacterial Medium (MBM) with 0.5% of peptone (Neutralised Bacteriological Peptone, Oxoid Ltd, Basingstoke, UK) in sterile filtered (Whatman 1001-270, pore size 11 μm) natural seawater. A spectroscopy was used to monitor bacterial growth at 630 nm. The starting concentration for bioassays of bacterial suspensions was 2 $\times$ 10^8 colony forming units/mL [21].

4.2 Vivatome Protocol

Zeiss Vivatome is a fluorescence microscopy that helps to collect necessary data for calculating an optical section in a single image. It shows images containing samples and bacteria and for each net sample, two images are recorded: the first image presents only fibers (brightfield) and the second image shows bacteria (green).

Bacteria colonization, on grafted and ungrafted nets, was determined using epifluorescence microscope VivaTome. First of all, *P. citrea* solution was distributed in 6 tubes containing PA 6.6 and HDPE samples (between 0.7 and 1 cm): one tube was for the ungrafted PA 6.6 net, one tube was for the ungrafted HDPE net, two tubes were for the PA 6.6 nets grafted with PEG, two tubes were for the HDPE nets grafted with PEG.

In order to study the antibacterial activity, samples were left covered at 20 °C, under stirring for 14 days, and they were collected after 1, 7, and 14 days. To fix the surface-attached bacteria, collected samples were placed in 3% formol for 10 min. Then, fibers from the tested samples were placed in sterile seawater and vortexed for 5 times to remove non-attached bacteria. Fibers were putted in Sybr Green Fluorescent DNA Stain for 5 min and finally examined with VivaTome (FITC filter, 10× objective lens).

4.3 *Image Processing with ImageJ*

ImageJ is a Java-based image processing program which can edit, analyze, and process images. It can also calculate area, measure distances, and angles and provide many other operations.

So, the obtained images, with Zeiss Vivatome, were processed with ImageJ in order to calculate the quantity of adhered bacteria on net surfaces. This amount presented the percentage of the surface-attached bacteria and calculated using Eq. (1).

$$\%\text{ adhered bacteria} = \frac{\text{Area occupied by bacteria}}{\text{Total area}} \times 100 \quad (1)$$

Then, and in order to determine the percentage change (increase or decrease) of bacteria adhesion on nets over time, the average percentage was calculated following Eq. (2):

$$\text{Average percentage }(\%) = \left| \frac{\text{first day adhered bacteria }\% - \text{last day adhered bacteria }\%}{\text{first day adhered bacteria }\%} \right| \times 100 \quad (2)$$

5 Results and Discussion

5.1 *Surface Topography Analyses*

Surfaces of grafted nets with PEG and chitosan were observed using a Scanning Electron Microscope (SEM) at a magnification of ×1000. The obtained images for PA 6.6 and HDPE nets showed different morphology changes after PEG grafting (Figs. 2 and 3).

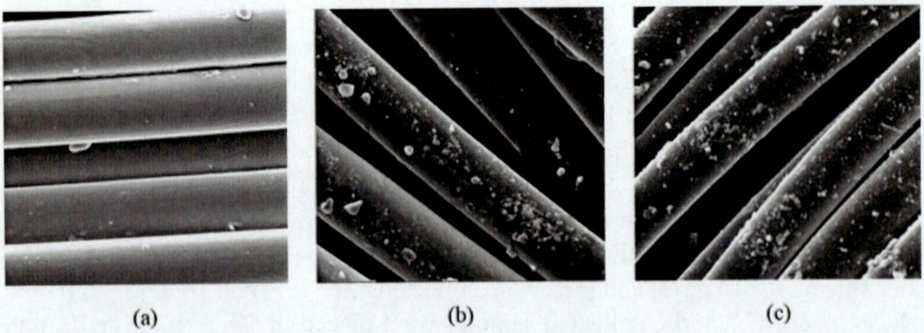

Fig. 2 SEM images at a magnification of 1000× of **a** ungrafted PA 6.6 net, **b** PA-Dop-PEG, **c** PA-Dop45°-PEG

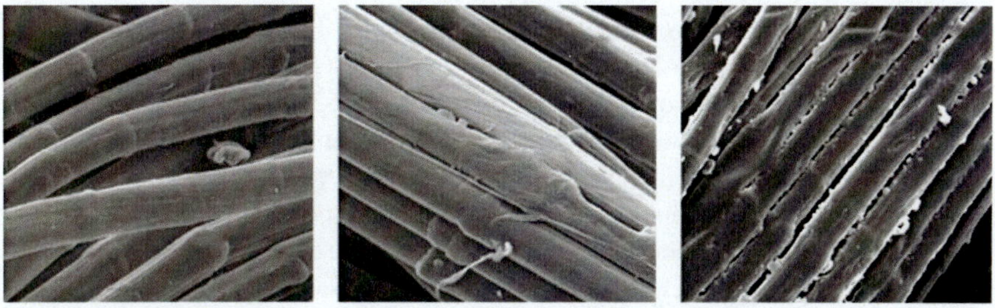

Fig. 3 SEM images at a magnification of 1000× of **a** ungrafted HDPE net, **b** HDPE-Dop-PEG, **c** HDPE-Dop45°-PEG

For PA 6.6 nets, new regular nodules appeared on PA 6.6 grafted nets, compared to the ungrafted nets. So PEG exists obviously on the surface, but they were poorly spread, leading to an irregular layer forming, for both grafting methods.

The surface topography of grafted HDPE nets was significantly rougher than that of the untreated nets, which might be explained by the presence of new groups on the surface.

This morphological study didn't show a significant result. So, an XPS analyze was mandatory in order to compare the chemical compositions of samples.

5.2 *X-ray Photoelectron Spectroscopy Results*

XPS is applied to examine the chemical composition and determine the exact quantity for each atom present in the composition of nets [2].

5.2.1 XPS Results for PA 6.6 Nets Grafted with PEG

Table 1 shows the atomic concentration data of ungrafted PA 6.6 nets and PA 6.6 nets grafted with PEG, measured by XPS.

Table 1 Surface composition of ungrafted and grafted PA 6.6 nets with PEG

PA 6.6 samples	Atomic percentage (%)			
	C	O	N	Si
Ungrafted PA 6.6	77.5	15.4	5.1	2
PA-Dop-PEG	69.9	26.6	1.7	1.5
PA-Dop45°-PEG	66.9	31.9	0.9	0

As mentioned before, the theoretical composition of PA 6.6 is 75.0 at% of C, 12.5 at% of O, and 12.5 at% of N. However, in this study, the untreated nets of PA 6.6 was composed of 77.5% of carbon, 15.4% of oxygen, 5.1% of nitrogen, and 2% of silicon. The difference between both compositions can be explained by the fact that the used nets have been dyed before as an industrial finishing.

For polyamide nets functionalized with dopamine at room temperature and then grafted with PEG, a relative decrease in carbon C and nitrogen N concentration and an increase in oxygen O, were noticed. The used PEG is composed of C (66.66%) and O (33.33%). That is why there is a remarkable decrease in the percentage of nitrogen N, N/O, and N/C ratios after grafting. These results confirm the successful immobilization of the PEG chains on the surface of the net [15, 23].

The PA net functionalized with dopamine at 45 °C and then grafted with PEG gave a better result compared to the other functionalization. However, for both cases, nitrogen still remains visible on the surface. This can be explained by the fact that the grafted PEG layers were thinner than the XPS sampling depth (10 nm) which was reported by Sagle et al. [20] (A 2 μm thick layer of poly (ethylene glycol) on a polyamide surface always results in a significant nitrogen content). However, X-ray photoelectron spectroscopy provides qualitative evidence for the presence of PEG on the surface [23].

However, when C 1s high resolution (Fig. 4) was considered, the successful of polyethylene glycol grafting on PA 6.6 surfaces was evidenced by a decrease of C–C links (285.0 eV) and the increase of C–O/C–OH (286.5 eV).

C 1s spectrums showed an increase of C–O bonds from 3.42% to 20.07% and 51.02%, respectively for PEG grafting after dopamine activation at room temperature (22 °C) and at 45 °C. And this increase indicated the presence of PEG since its structure is composed by C–O linkages (Fig. 5). Campelo et al. observed similar behavior in their grafting study with polyethylene glycol [9]. Besides, the functionalization at 45 °C showed a better grafting of PEG. The polyethylene glycol grafting mechanism on PA 6.6 nets is presented in Fig. 5.

Since PA 6.6 grafted with PEG after activation at 45 °C showed a better grafting result, those nets were selected to test their antibacterial activity.

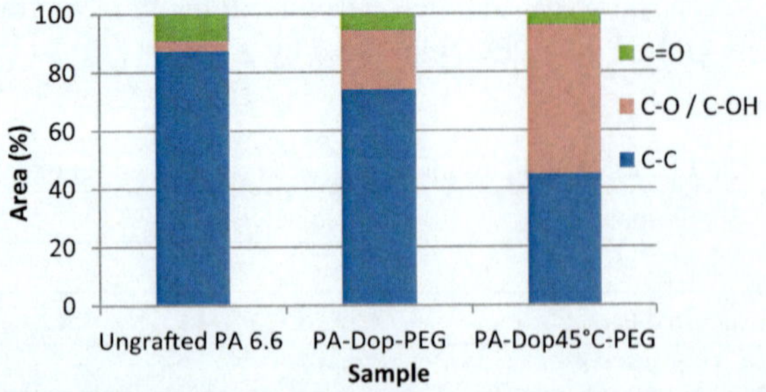

Fig. 4 Area (%) of different C 1s components from XPS analysis of ungrafted and grafted PA 6.6 with PEG (C–C: 285 eV; C–O/C–OH: 286 eV; C=O: 288 eV)

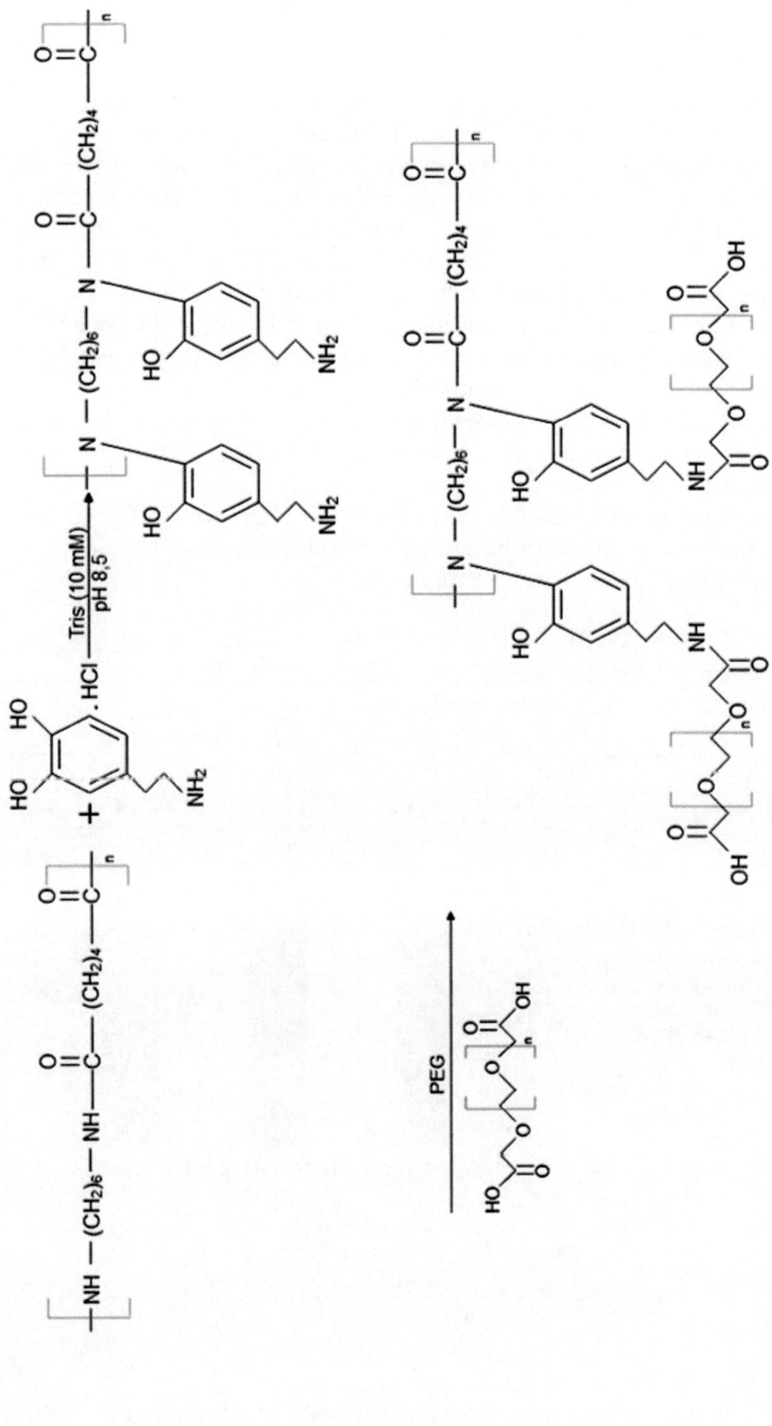

Fig. 5 Mechanism of PEG grafting on polyamide 6.6 nets

5.2.2 XPS Results for HDPE Nets Grafted with PEG

Table 2 shows the atomic concentration data of ungrafted HDPE net and HDPE nets grafted with PEG, obtained with XPS.

XPS showed that the unmodified HDPE is composed of 96.4% of C and 2.8% of O, and this composition depends of the followed industrial finishing process.

For HDPE nets grafted with PEG, a relative decrease in carbon concentration and an increase in oxygen concentration were noticed. The appearance of nitrogen atoms N is explained by the fact that the nets have been first functionalized with dopamine which consists of 72.7% of carbon, 18.2% of oxygen, and 9.1% of nitrogen. Besides, the HDPE net functionalized with dopamine at 45 °C favored a better result compared to the net functionalized at ambient temperature, because the composition of PEG is 66.66 at% C and 33.33 at% O. However, in both cases, traces of the nitrogen N (of the dopamine) still remain visible on the surface.

Moreover, C 1s bonds (Fig. 6) showed the appearance of new peaks at around 287 eV and 289 eV and they present respectively C–O and O=C–O bonds related to the PEG grafting (Fig. 7).

HDPE nets grafted with PEG after activation at 45 °C showed a better grafting result. So, they were selected to be analyzed at laboratory scale to study their antibacterial activity against three bacteria strains.

Table 2 Surface composition of ungrafted and grafted HDPE nets with PEG

HDPE samples	Atomic percentage (%)			
	C	O	N	Si
Ungrafted HDPE	96.4	2.8	0	0.8
HDPE-Dop-PEG	83.9	12.9	2.7	0.5
HDPE-Dop45°-PEG	80.6	15.8	2.3	1.3

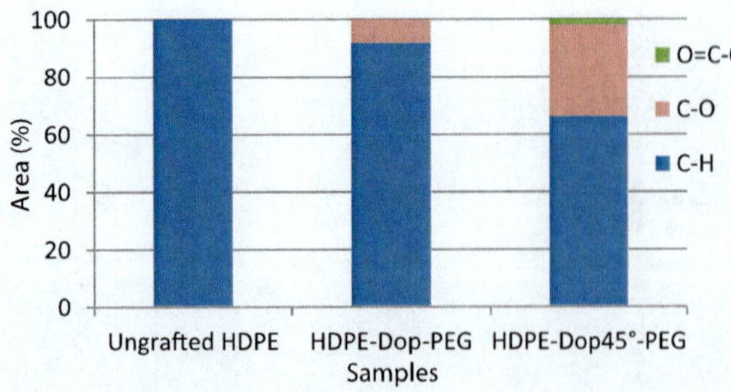

Fig. 6 Area (%) of different C 1s components from XPS analysis of ungrafted and grafted HDPE with PEG (C–H: 285 eV; C–O: 287 eV; O–C=O: 288 eV)

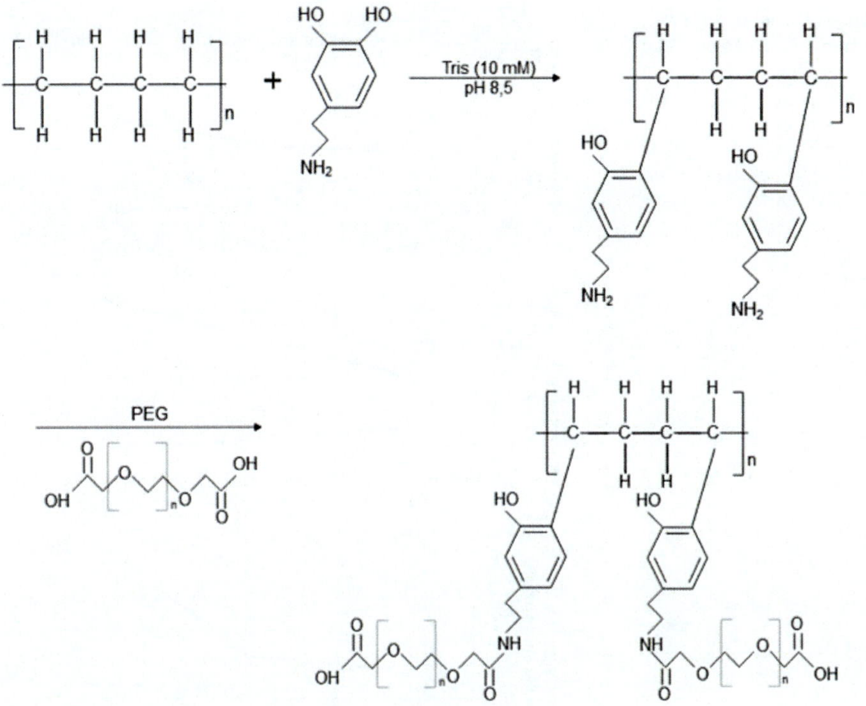

Fig. 7 Grafting mechanism of PEG on HDPE nets

5.3 *Antibacterial Behavior of Grafted PA 6.6 Nets*

The inhibition effect of PA 6.6 samples was tested on *P. citrea, P. elyakovii,* and *V. harveyi,* for a period of 14 days. For this, the method mentioned above (part 4) was followed. The test period was selected because the formation of marine biofilm as well as its maturation can be achieved within 24 h and even can reach a few days, depending on the strains and the environment [3, 8].

After impregnation in the bacterial solutions, ungrafted, and grafted PA 6.6 nets were analyzed with Vivatome and imageJ software. Then, the quantity of bacteria adhered to the surface of all samples was calculated using Eq. (1), and results are presented in Fig. 8.

The ungrafted PA 6.6 net showed an increase on the bacteria adhesion of 42.7%, 72.5%, and 201.4%, respectively for *P. citrea, P. elyakovii,* and *V. harveyi*. However, the PA 6.6 nets, functionalized with dopamine at 45 °C and then grafted with PEG, showed a good antibacterial activity with a decrease of bacteria adhesion' percentage equal to 80.5%, 38.4%, and 31.8%, respectively for *P. citrea, P. elyakovii,* and *V. harveyi*.

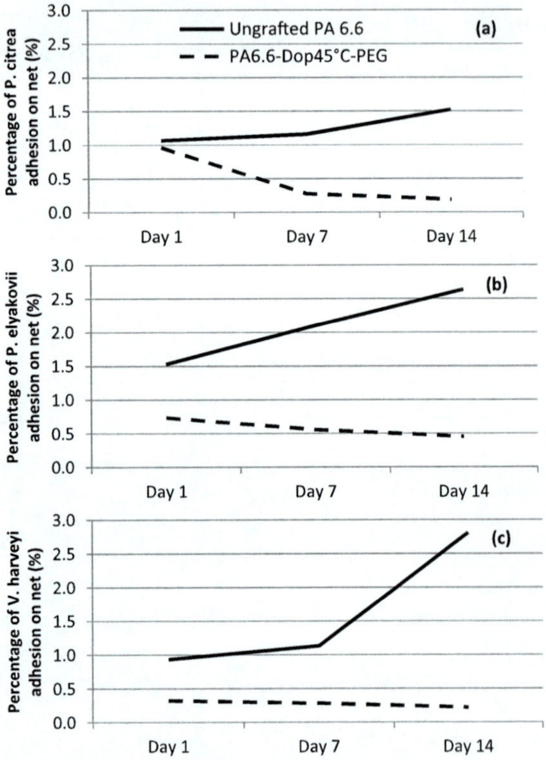

Fig. 8 Evolution of **a** *P. citrea*, **b** *P. elyakovii*, and **c** *V. harveyi* adhesion on untreated and grafted PA 6.6 over 2 weeks

5.4 Antibacterial Behavior of Grafted HDPE Nets

The evolution of bacteria adhesion on ungrafted and grafted HDPE nets is shown in Fig. 9.

Figure 9 shows that the percentage of bacteria adhered to the surface of untreated HDPE evolves in an increasing order with an average percentage of 73.8%, 39.5%, and 63.7%, respectively for *P. citrea*, *P. elyakovii,* and *V. harveyi.*

For the HDPE net functionalized with dopamine at 45 °C and grafted with PEG, bacterial adhesion was remarkably decreased with a percentage of 12.3%, 33.7%, and 39.1%, respectively for *P. citrea*, *P. elyakovii,* and *V. harveyi.*

Those results proved that PEG inhibited the bacteria surrounding the grafted PA 6.6 and HDPE nets, which explains the reduction of bacterial adhesion over time.

Hamming et al. [14] also found that grafted PEG coatings can provide effective resistance to fouling of surfaces by proteins, bacteria, and other biological organisms. Therefore, those coatings can be used to control fouling of ship hulls, implantable medical devices, surgical tools, etc.

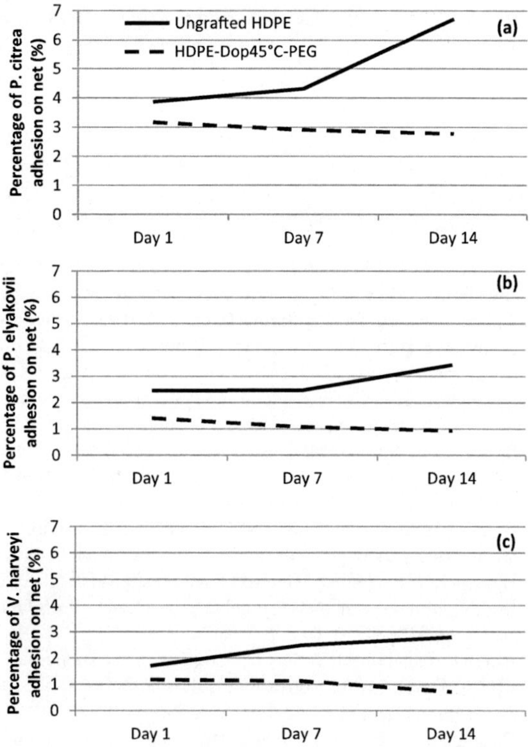

Fig. 9 Evolution of **a** *P. citrea*, **b** *P. elyakovii,* and **c** *V. harveyi* adhesion on untreated and grafted HDPE over 2 weeks

6 Conclusion

This study was about finding an antibacterial solution in order to limit the adhesion and formation of biofilm, and subsequently, reducing the biofouling phenomenon. That's why, the research aims are to find new solution to inhibit biofouling and to develop new environmentally friendly solutions which limit bacteria adhesion on aquaculture nets.

This study showed that the grafting of PEG after functionalization at 45 °C favored a better result and a better fixation of PEG on both PA 6.6 and HDPE nets. Also, this grafting showed a good antibacterial behavior toward the three bacteria strains *P. citrea, P. elyakovii,* and *V. harveyi*. Indeed, grafted PA 6.6 nets showed a decrease of bacteria adhesion equal to 80.5%, 38.4%, and 31.8%, respectively for *P. citrea, P. elyakovii,* and *V. harveyi*. Moreover, grafted HDPE nets showed a decrease in the percentage of adhered bacteria of 12.3%, 33.7%, and 39.1%, respectively for *P. citrea, P. elyakovii,* and *V. harveyi*.

These results may confirm the effectiveness of PEG grafting and its resistance to bacteria and proteins adhesion. Also, they can promise an attractive industrial solution that may limit aquatic biofouling on aquaculture nets. However, it is mandatory to study the toxicity of PEG on algae, crustaceans, fishes, and other macro marine organisms in order to know its effects on aquatic environment. Also, a fastness evaluation must be realized to know the lifetime and stability of the grafted nets.

References

1. Amara I, Miled W, BenSlama R et al (2018) Antifouling processes and toxicity effects of antifouling paints on marine environment A review. Environ Toxicol Pharmacol 57:115–130
2. Amara I, Miled W, BenSlama R et al (2019) Surface modifications by plasma treatment, chemical grafting and over dyeing of polyamide nets to improve the antifouling performance in the aquaculture field. Dyes Pigm 166:107–113. https://doi.org/10.1016/j.dyepig.2019.03.005
3. Aye AM (2015) Mise en évidence du système de communication "Quorum Sensing" impliquant les AHLs chez des bactéries marines isolées de la Méditerranée, Toulon
4. Bannister J, Sievers M, Bush F, et al (2019) Biofouling in marine aquaculture: a review of recent research and developments. Biofouling 35(6):631–648. https://doi.org/10.1080/08927014.2019.1640214
5. Baxter EJ, Sturt MM, Ruane NM, et al (2012) Biofouling of the hydroid Ectopleura larynx on aquaculture nets in Ireland: implications for finfish health. Fish Vet J
6. Bloecher N, Powell M, Hytterød S, et al (2018) Effects of cnidarian biofouling on salmon gill health and development of amoebic gill disease. PLoS One 13(7). https://doi.org/10.1371/journal.pone.0199842
7. Bovio E, Fauchon M, Toueix Y et al (2019) The sponge-associated fungus Eurotium chevalieri MUT 2316 and its bioactive molecules: potential applications in the field of antifouling. Mar Biotechnol 21(6):743–752. https://doi.org/10.1007/s10126-019-09920-y
8. Brian-Jaisson F (2014) Identification et caractérisation des exopolymères de biofilms de bactéries marines, Toulon
9. Campelo CS, Chevallier P, Vaz JM et al (2017) Sulfonated chitosan and dopamine based coatings for metallic implants in contact with blood. Mater Sci Eng C 72:682–691
10. Dunne WM (2002) Bacterial adhesion: seen any good biofilms lately? Clin Microbiol Rev 15(2):155–166. https://doi.org/10.1128/CMR.15.2.155-166.2002
11. Eckman JE, Thistle D, Burnett WC et al (2001) Performance of cages as large animal-exclusion devices in the deep sea. J Mar Res 59(1):79–95. https://doi.org/10.1357/002224001321237371
12. Guardiola FA, Cuesta A, Meseguer J et al (2012) Risks of using antifouling biocides in aquaculture. Int J Mol Sci 13(2):1541–1560. https://doi.org/10.3390/ijms13021541
13. Hall-Stoodley L, Stoodley P (2002) Developmental regulation of microbial biofilms. Curr Opin Biotechnol 13(3):228–233. https://doi.org/10.1016/S0958-1669(02)00318-X
14. Hamming LM, Messersmith PB (2008) Fouling resistant biomimetic poly (ethylene glycol) based grafted polymer coatings. Mater Matters 3(52)
15. Kang G, Liu M, Lin B et al (2007) A novel method of surface modification on thin-film composite reverse osmosis membrane by grafting poly(ethylene glycol). Polymer 48(5):1165–1170
16. Lane A, Willemsen P (2004) Collaborative effort looks into biofouling. Fish Farming Int 44:34–35
17. Lehaitre M, Delauney L, Compère C (2008) Biofouling and underwater measurements. Real-time observation systems for ecosystem dynamics and harmful algal blooms: theory, instrumentation and modelling. Oceanographic methodology series. UNESCO, Paris, pp 463–493
18. Li B and Ye Q (2015). Antifouling surfaces of self-assembled thin layer. In: Antifouling surfaces and materials. Springer, pp 31–54
19. Phillippi AL, O'Connor NJ, Lewis AF et al (2001) Surface flocking as a possible anti-biofoulant. Aquac 195(3–4):225–238. https://doi.org/10.1016/S0044-8486(00)00556-1

20. Sagle AC, Van Wagner EM, Ju H, et al (2009) PEG-coated reverse osmosis membranes: desalination properties and fouling resistance. J Membr Sci 340(1–2):92–108
21. Trepos R, Cervin G, Pile C et al (2015) Evaluation of cationic micropeptides derived from the innate immune system as inhibitors of marine biofouling. Biofouling 31(4):393–403. https://doi.org/10.1080/08927014.2015.1048238
22. Vaz JM, Michel EC, Chevallier P et al (2014) Covalent grafting of chitosan on plasma-treated polytetrafluoroethylene surfaces for biomedical applications. J Biomater Tissue Eng 4(11):915–924
23. Van Wagner EM, Sagle AC, Sharma MM et al (2011) Surface modification of commercial polyamide desalination membranes using poly(ethylene glycol) diglycidyl ether to enhance membrane fouling resistance. J Membr Sci 367(1–2):273–287
24. Walker J, Marsh P (2004) A review of biofilms and their role in microbial contamination of dental unit water systems (DUWS). Int Biodeterior Biodegrad 54(2–3):87–98. https://doi.org/10.1016/j.ibiod.2004.03.012
25. Willemsen P (2005) Biofouling in European aquaculture: is there an easy solution. Eur Aquac Soc 35:82–87
26. Yebra DM, Kiil S, Dam-Johansen K (2004) Antifouling technology—past, present and future steps towards efficient and environmentally friendly antifouling coatings. Prog Org Coat 50(2):75–104. https://doi.org/10.1016/j.porgcoat.2003.06.00

Ecofriendly Extraction of Natural Dye from Waste of Cucurbita Pepo Using Assisted Microwave Method

Marwa Souissi, Hana Hania, Nizar Meksi, and Hatem Dhaouadi

Abstract This paper introduces the feasibility of extraction a natural dye from the waste of Cucurbita pepo. The extraction was carried out with an unconventional ecofriendly method of extraction which consists of the use of an assisted microwave. The chemical composition of extracted solution from the waste of Cucurbita pepo is characterized in terms of total polyphenols, condensed tannins, chlorophyll, and carotenoids. The evaluated parameters are the dry mass of plant matter, the power of assisted microwave, and the duration of extraction. Obtained results were analyzed in order to deduce the optimal conditions for extraction natural dye from the waste of Cucurbita pepo. Obtained results indicate that the optimal extraction conditions are 70 g/L of dry mass of plant matter, 750 W of microwave irradiation power, and extraction time of 20 min.

1 Introduction

Nowadays, because of the awareness of the huge danger caused by the use of synthetic dyes and their additives, natural dyes seem to be the best ecofriendly solution [1, 2]. Natural dyes are currently arousing renewed interest [3, 4]. In this context, the extraction of natural dyes has resurfaced using conventional extraction methods such as decoction, infusion, maceration, percolation and pressing or unconventional processes such as ultrasonic extraction, supercritical fluid extraction, and microwave extraction [3, 5].

Extraction with supercritical fluids [5] is a very attractive method that deserves special attention. The advantages of this method over conventional methods are shorter extraction time, high selectivity, and ease of solvent removal after extraction

M. Souissi (✉) · N. Meksi · H. Dhaouadi
Laboratory of Environmental Chemistry and Cleaner Process (LCE2P-LR21ES04), University of Monastir, 5019 Monastir, Tunisia
e-mail: souissi.marwa20@yahoo.com

M. Souissi · H. Hania · N. Meksi
National Engineering School of Monastir, University of Monastir, 5019 Monastir, Tunisia

R. Khiari and M. Jawaid (eds.), *Proceedings of the 3rd International Congress of Applied Chemistry & Environment (ICACE–3)*, Springer Proceedings in Materials 23, https://doi.org/10.1007/978-981-99-1968-0_2

Fig. 1 Preparation of plant material

by simple decompression. However, it remains an expensive technique that requires extreme operating conditions [6].

Likewise, microwave assisted extraction (MAE) is another alternative process that has recently received a lot of attention in the food industry since this technology allows meeting industrial needs dictated by environmental regulations and has many advantages over traditional extraction methods, such as higher extraction rates with lower costs, and time savings. Microwave heating is done from the inside out [5]. This particularity leads to an extraction of the plant active ingredients by internal overpressure and osmotic effect, which allows a very efficient, very fast, solvent-free extraction, and a high concentration of the extracts [7]. Microwave extraction is then an environmentally friendly extraction technique. The present study focuses on extraction of a natural dye from waste of Cucurbita Pepo to be used to dye PET filaments.

The microwave irradiation power, the extraction time, and the concentration of plant material are the main experimental parameters that affect the extraction operation. The influence of these parameters on absorbance, yields of polyphenols, and condensed tannins will be studied.

2 Materials and Methods

2.1 Preparation of Plant Material

For the preparation of Cucurbita pepo, wastes were cut into fine strips, dried in an oven at 80 °C for 24 h, and then finely ground to be transformed into powder (Fig. 1).

2.2 Aqueous Extraction of Coloring Matter

The extraction technique used consists of impregnating a mass of coloring matter in 250 mL of water. The whole was placed in a reflux assembly placed in an assisted microwave. The mass of coloring matter, the microwave irradiation power, and the duration of extraction are the three parameters to be studied and optimized in the case of natural dyes when dyeing 100% PET filaments.

2.3 *Determination of Polyphenols Content*

This assay is based on the method using the Folin–Ciocalteu reagent [3]. This reagent consists of a mixture of phosphomolybdic and phosphotungstic acids [8]. All details of this assay have been mentioned in our previous work [3].

2.4 *Determination of Condensed Tannins Content*

Condensed tannins content was estimated using the Julkunen-Titto method [9]. A volume of 400 μL of the extract was added to 3 mL of a 4% methanolic solution of vanillin. Then, a volume of 1.5 mL of concentrated hydrochloric acid was added. After 15 min of reaction, the absorbance was measured at wavelength of 550 nm.

2.5 *Determination of Carotenoids and Chlorophyll Contents*

The determination of the amounts of these pigments has been described by Lichtenthaler [10]. In fact, in a 25 mL volumetric flask, 0.21 g of our coloring matter powder was soaked in acetone and mixed for 30 min at 37 °C. We passed the contents in the centrifuge at room temperature for 10 min and at 3500 rpm to obtain a clear solution. Then, we determined the content of chlorophylls and carotenoids after measuring the absorbance.

For all the secondary metabolite assays studied, each test was carried out three times and the corresponding result was expressed as an arithmetic mean. The experimental errors were estimated using the measurement of the standard deviations.

3 Results and Discussions

3.1 *Study of the Effect of the Concentration of Coloring Matter*

The effect of varying the concentration of coloring matter on the absorbance is illustrated in Fig. 2. Obtained results showed that from 70 g/L, the absorbance value is maximum. Above this concentration, the absorbance value remains constant.

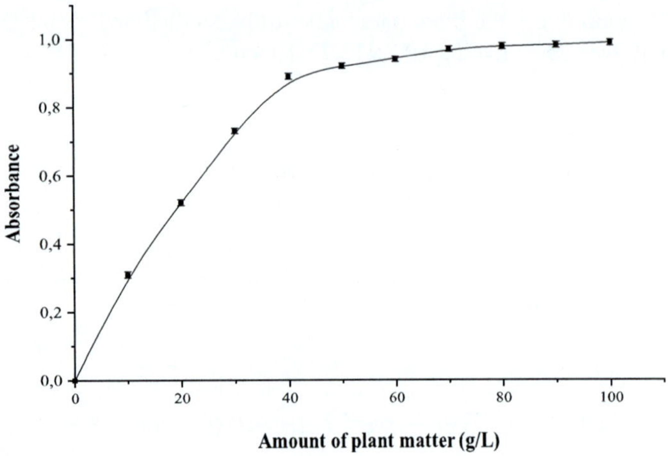

Fig. 2 Effect of the amount of waste Cucurbita pepo dry matter on absorbance

3.2 Extraction Power Effect

The influence of the power of assisted microwave on the absorbance of the staining solution is shown in Fig. 3. From Fig. 3, it can be observed a positive evolution of the absorbance according to the increase in microwave power to reach a value of the order of 0.91 for a maximum power equal to 750 W.

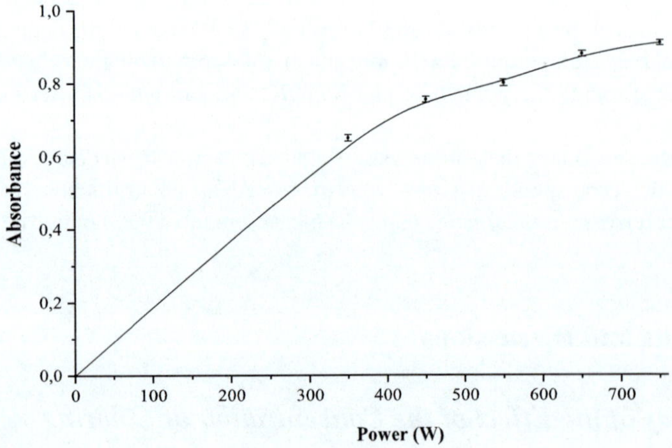

Fig. 3 Effect of microwave irradiation power (W) on absorbance

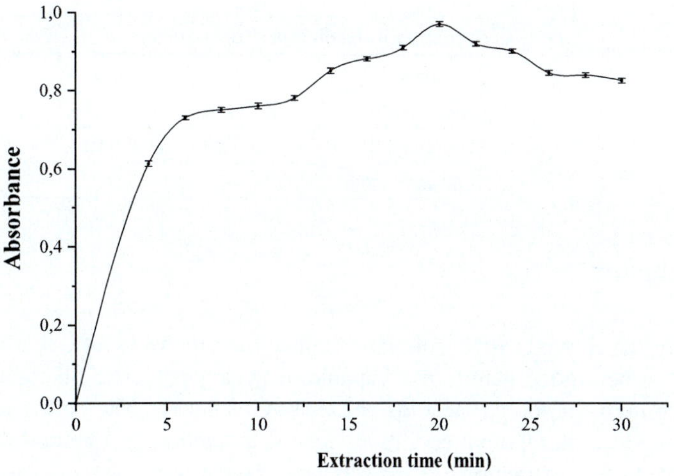

Fig. 4 Effect of extraction time on absorbance

3.3 Study of the Effect of Extraction Duration

The effect of extraction time on absorbance is illustrated in Fig. 4. As can be seen, the absorbance increases with increasing extraction time and reaches the maximum value of 0.97 for an extraction time of 20 min. Beyond this value, a degradation of the coloring solution is observed.

3.4 Content of Secondary Metabolites of Cucurbita Pepo

The coloring solution extracted from the waste of Cucurbita pepo under optimal extraction conditions allowed us to determine its chemical composition in terms of phenolic compounds, chlorophyll, and carotenoids. Obtained results are summarized in Table 1. It can be seen that Cucurbita pepo contains a low percentage of total polyphenols and chlorophyll compared to the content of carotenoids. The high carotenoids value proves that this natural dye is suitable for dyeing polyester fibers because it has been observed in the literature that compounds containing carotenoids have a high affinity with hydrophobic and difficult-to-dye polyester fibers [6].

Table 1 Content of secondary metabolites of Cucurbita pepo

Secondary metabolites tested	Content (mg/g of dry matter)
Total polyphenols	6.43 ± 0.02
Chlorophyll a	1.732 ± 0.01
Chlorophyll b	1.738 ± 0.02
Total carotenoïds	37.19 ± 0.23

4 Conclusion

In this study, the colored extract solution obtained from the waste of Cucurbita pepo was found to be a good natural dye capable of dyeing polyester fibers due to its chemical composition on carotenoids and total polyphenols. The objective of this work is to deduce the optimal conditions for the extraction of a natural dye from Cucurbita pepo. The extraction process was carried out using an assisted microwave. Obtained results showed that optimum conditions are 70 g/L of dry mass of plant matter, a microwave irradiation power of 750 W, and extraction time of 20 min.

References

1. Souissi M, Khiari R, Zaag M, Meksi N, Dhaouadi H (2021) Ecological and cleaner process for dyeing bicomponent polyester filaments (PET/PTT) using ecological carriers: analysis of dyeing performance. RSC Adv 11:25830–25840
2. Souissi M, Khiari R, Zaag M, Meksi N, Dhaouadi H (2022) Kinetic study of dyeing bicomponent polyester textiles (PET/PTT) using environmentally friendly carriers. RSC Adv 12:2361–2374
3. Souissi M, Guesmi A, Moussa A (2018) Valorization of natural dye extracted from date palm pits (Phoenix dactylifera) for dyeing of cotton fabric. Part 1: optimization of extraction process using Taguchi design. J Clean Prod 202:1045–1055
4. Souissi M, Guesmi A, Moussa A (2018) Valorization of natural dye extracted from date palm pits (Phoenix dactylifera) for dyeing of cotton fabric. Part 2: optimization of dyeing process and improvement of colorfastness with biological mordants. J Clean Prod 204:1143–1153
5. Krishnananda PI, Amit GD, Dipika AP, Mahendra SD, Mangesh PM, Vaibhav CK (2017) Phytochemicals: extraction methods, identification and detection of bioactive compounds from plant extracts. J Pharmacogn Phytochem 1:32–36
6. Abate MT, Ferri A, Guan J, Chen G, Nierstrasz V (2019) Colouration and bio-activation of polyester fabric with curcumin in supercritical CO_2: Part I—investigating colouration properties. J Supercrit Fluids
7. Chungkrang L, Bhuyan S, Phukan AR (2020) Natural dye sources and its applications in textiles: a brief review. Int J Curr Microbiol App Sci 10:261–269
8. Khasnabis J, Rai C, Roy A (2015) Determination of tannin content by titrimetric method from different types of tea. J Chem Pharm 6:238–241
9. Das BK, Al-Amin MM, Russel SM, Kabir S, Bhattacherjee R, Hannan JMA (2014) Phytochemical screening and evaluation of analgesic activity of Oroxylum indicum. Indian J Pharm Sci 571–575
10. Lichtenthaler HK, Buschmann C (2001) Chlorophylls and carotenoids-extraction, isolation and purification, current protocols in food analytical chemistry (CPFA). Hoboken, New Jersey

Adsorption Studies of Atrazine Herbicide from Aqueous Solutions Using *Typha australis* as an Inexpensive Adsorbent: A Preliminary Investigation

Abdoulaye Demba N'Diaye, Noureddine Baaka, Mohamed Kankou, and Hatem Dhaouadi

Abstract The present work has been performed to explore the employability of the powder of *Typha australis* leaves as low cost adsorbent for the adsorption of atrazine (ATZ) herbicide in batch procedures and continuous column operations. Pseudo-first order (PFO) and pseudo-second order (PSO) kinetic models with nonlinear methods were applied. The experimental data is perfectly adapted to the PSO model. Equilibrium data were described using Langmuir, Freundlich, and Redlich–Peterson isotherm models with a nonlinear approach. The Redlich–Peterson isotherm and Pearson method confirmed the suitability of Langmuir. Based on isotherm parameter determination, *Typha australis* leaves adsorbent capacity of 0.87 mg g^{-1} was obtained. Regarding the column study, the results confirm that the breakout curve is dependent on the initial ATZ concentration with a peak laminar volume of 1.37 mg g^{-1} at an initial ATZ concentration of 20 mg L^{-1} and the flow rate is 2 mL min^{-1}.

Keywords Atrazine · Herbicide · *Typha australis* · Adsorbent

1 Introduction

The herbicide atrazine (ATZ) was distinguished by its continuous journey in soil and higher water leaching potential. ATZ is frequently found in aquatic environments, due to its widespread usage [1, 2].

A. D. N'Diaye (✉) · N. Baaka · H. Dhaouadi
Faculty of Sciences of Monastir, Research Laboratory—Environmental Chemistry and Clean Processes (LR21ES04), University of Monastir, 5000 Monastir, Tunisia
e-mail: abdouldemba@yahoo.fr

A. D. N'Diaye · M. Kankou
UR Eaux Pollution et Environnement, Département de Chimie, Faculté des Sciences et Techniques, Université de Nouakchott Al Aasriya, Nouakchott, Mauritanie

R. Khiari and M. Jawaid (eds.), *Proceedings of the 3rd International Congress of Applied Chemistry & Environment (ICACE-3)*, Springer Proceedings in Materials 23, https://doi.org/10.1007/978-981-99-1968-0_3

However, ATZ could pose a potential risk to humans [3] and animals [4]. Its use has been forbidden in European Union countries [5]. This herbicide is still used in China. European Union legislation permits very low ATZ concentrations in drinking water, while China sets this limit at 3 ppb [5]. ATZ contamination of surface and groundwater has increased public concern about its environmental impact. ATZ has recently been recognized as an endocrine disruptor in mammals [6] and aquatic organisms [7].

Several processes have been researched and developed to remove ATZ, including chemical oxidation [8], photocatalysis [9], photocatalysis and biological degradation [10], Fenton degradation [11], membrane elimination [12], electrochemical [13], and adsorption [14].

However, among these processes, the adsorption technique is most effective for the elimination of pesticides from aquatic media [15].

Additionally, the treatment of water has historically made extensive use of activated carbon. However, the price of the activated carbon with adsorbent grade is exorbitant. For this reason, we are motivated to develop natural alternatives to the conventional carbon adsorbent for the treatment of water that are low cost and readily available locally.

To address this issue, the use of less expensive and indigenous waste materials for pesticide removal from aqueous solutions has become a focus of significant research. Many materials have been used for ATZ removal such as press-cake from oil seeds [16], Zeolite-A Prepared from Egyptian kaolin [17, 18] banana peel, Moringa Oleifera Pods [19].

In this work *Typha australis* was used to examine its ability to sorb ATZ in solution. *Typha australis* was selected as it is a widespread and dominant plant in many aquatic systems. So, the efficacy and feasibility for ATZ adsorption onto the batch experiments with the fixed bed column of *Typha australis* leaves was explored.

2 Experimental

2.1 *Adsorbate and Adsorbent*

Analytical reagent grade is used for all the study. Details on the preparation of the stock solution ATZ (1000 mg L^{-1}), as well as some dilutions have been reported by N'Diaye et al. [14] (Fig. 1).

The biomass was harvested in the Wilaya of Trarza, Mauritania. Details on the preparation and physicochemical parameters of the *Typha australis* leaves powder have been previously reported by N'diaye et al. [20]. The physicochemical parameters of the studied leaves of *Typha australis* are inclined in Table 1.

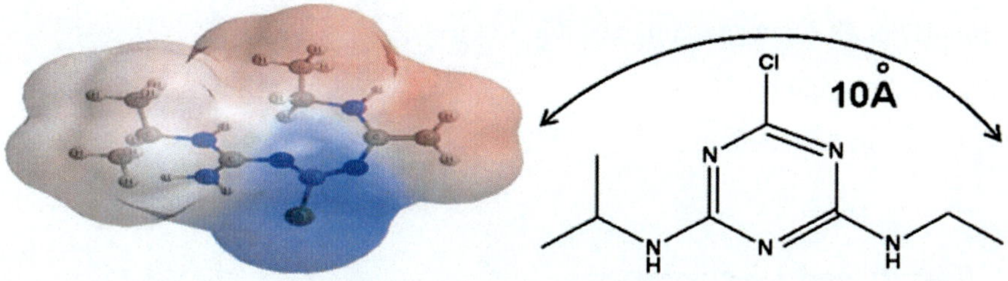

Fig. 1 ATZ chemical structure

Table 1 Results of the physicochemical study of *Typha australis* leaves adsorbent

Parameters	Values
Cellulose (%)	38.5
Hémicellulose (%)	27.8
Lignine (%)	14.9
C (%)	43.93
H (%)	5.87
N (%)	0.88
S (%)	0.28
O (%)	49.04
Particle size (μm)	<100

2.2 Batch Experiments

To investigate the effect of exposure time on ATZ removal, 0.5 g of adsorbent was added to 25 mL of initial ATZ concentration (5–10 mg L^{-1}). Discontinuous isothermal procedures were performed at different initial concentrations of ATZ (2.5–100 mg L^{-1}).

The residual concentrations of ATZ were analysed by High Performance Liquid of Chromatography. The adsorption capacity (q_e) and percentage of ATZ removed R (%) was determined using, respectively, following Eqs. (1) and (2):

$$q_e = \frac{(C_i - C_e)V}{m} \quad (1)$$

$$R\ (\%) = \frac{C_i - C_e}{C_i} \times 100 \quad (2)$$

where

q_e (mg g^{-1}): capacity ATZ concentration adsorbed by adsorbent,

C_i (mg L^{-1}): ATZ initial concentrations,

C_e (mg L^{-1}): ATZ concentration at equilibrium,

V (L): solution volume,

M (g): mass of adsorbent used.

3 Results and Discussion

3.1 Kinetic Study

Figure 2 shows the influence of contact time on the elimination of 5 and 10 mg L^{-1} ATZ by *Typha australis* leaves adsorbent. From Fig. 2, we can see that the adsorption was faster; the equilibrium is reached in about 40 min.

N'diaye et al. [14], also, found that the biosorbent equilibrium time was short.

Rates and mechanisms were calculated using, both, a pseudo-first order (PFO) model Eq. (3) [21] and a Pseudo-Second-Order model (PSO) Eq. (4):

$$q_t = q_e(1 - \exp^{-k_1 t}) \tag{3}$$

$$q_t = \frac{k_2 q_e^2 t}{1 + k_2 q_e t} \tag{4}$$

where

q_t (mg g^{-1}): adsorbed amount of ATZ adsorbed by adsorbent,

k_1 (L min^{-1}): PFO rate constant,

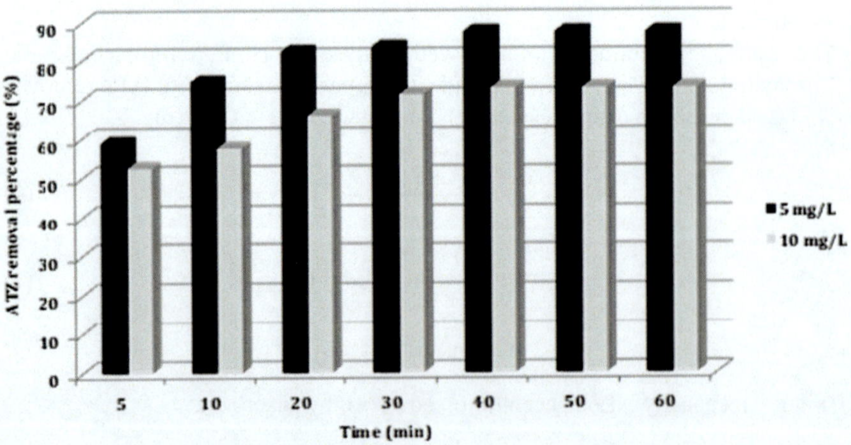

Fig. 2 ATZ removal (%) by *Typha australis* leaves as adsorbent

k_2 (g mg^{-1} min^{-1}): PSO rate constant,

q_e (mg g^{-1}): amount of ATZ adsorbed at equilibrium,

t (min): contact time.

From results of Figs. 3 and 4 and Table 2, it can be seen that the experimental data is perfectly adapted to the PSO model which indicated that the mechanism was controlled by chemisorption [22].

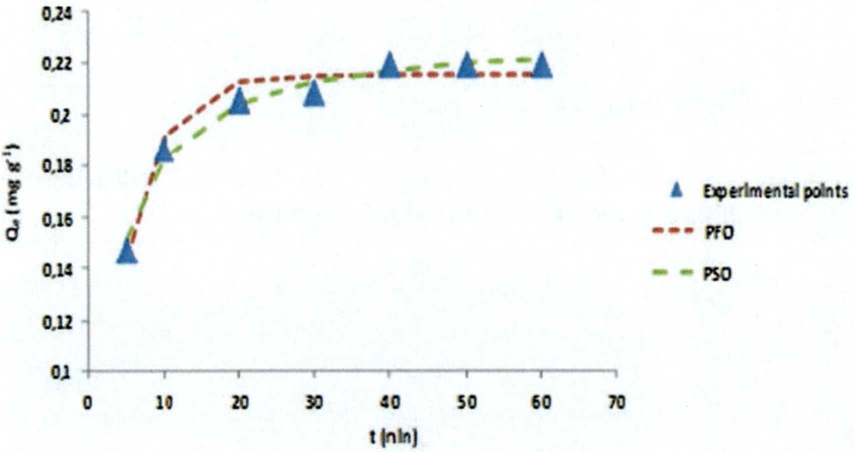

Fig. 3 Non-linearized kinetics by *Typha australis* leaves using 5 mg L^{-1} initial ATZ concentration

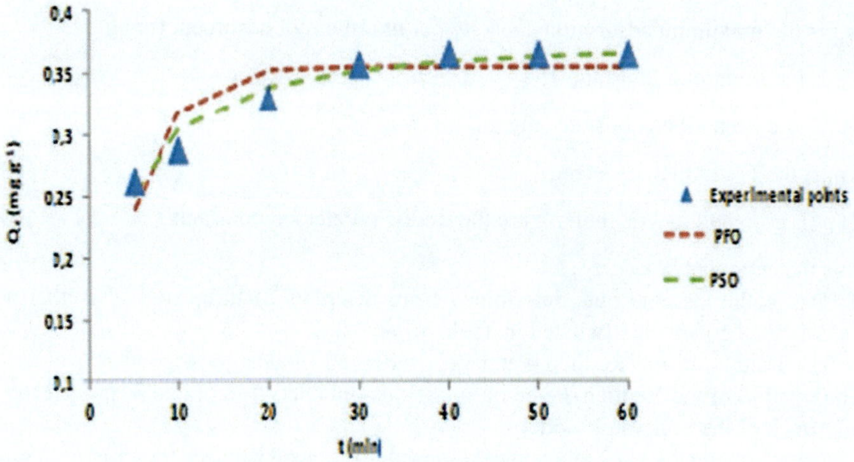

Fig. 4 Non-linearized kinetics by *Typha australis* leaves using 10 mg L^{-1} initial ATZ concentration

Table 2 R^2 and kinetic constants for adsorption of ATZ onto *Typha australis*

Model	Parameters	5 mg L^{-1}	10 mg L^{-1}
PFO	q_{exp}	0.219	0.366
	q_e	0.214	0.355
	k_1	0.221	0.225
PSO	R^2 (%)	99.4	89.9
	q_e	0.231	0.382
	k_2	1.61	1
	R^2 (%)	99.6	97.4

3.2 *Study of Adsorption Isotherms*

In order to study equilibrium adsorption, the Langmuir Eq. (5), Freundlich Eq. (6) [23], and Redlich–Peterson Eq. (7) [24] models have been used:

$$q_e = \frac{q_m K_L C_e}{1 + K_L C_e} \quad (5)$$

$$q_e = K_F C_e^{1/n} \quad (6)$$

$$q_e = \frac{K_{RP} C_e}{1 + \alpha_{RP} C_e^n} \quad (7)$$

where

q_m: is the maximum adsorption capacity per unit mass of adsorbent (mg g^{-1}),

k_L: is the Langmuir constant (L g^{-1}),

K_F: is the Freundlich constant (mg g^{-1}) (L $mg^{-1})^n$,

$1/n$: is the Freundlich constant,

K_{RP} (L g^{-1}) and α_{RP} (L mol^{-1}): are the Redlich–Peterson constants,

n: is the exponent, $0 < n < 1$.

The isotherms constants determined from the plots of three studied isotherm models (Fig. 5) are clearly listed in Table 3.

The Langmuir and Redlich–Peterson isothermal models gave good R^2 values. The n-value of the Redlich–Peterson model, as obtained from Table 3, proved the relevance of the Langmuir model.

On the other hand, the higher correlation value obtained between the experimental and modelled Langmuir data ($r = 0.987$) (Table 4) supported the fact that the experimental equilibrium data are perfectly matched to Langmuir. Based on isotherm parameter determination, *Typha australis* leaves adsorbent capacity of 0.87 mg g^{-1} was obtained.

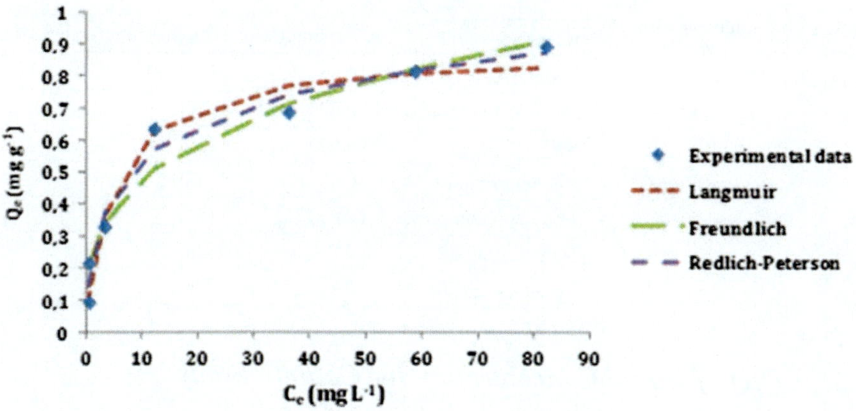

Fig. 5 Isotherm models for ATZ adsorption by *Typha australis*

Table 3 Results of ATZ adsorption study by *Typha australis*

Model	Parameters	Values
Langmuir	q_m	0.87
	K_L	0.21
	R^2 (%)	99.90
Freundlich	n	3.27
	K_F	0.24
	R^2 (%)	93.40
Redlich–Peterson	K_{RP}	0.36
	α_{RP}	0.85
	n	0.83
	R^2 (%)	99.97

Table 4 Correlation matrix of the isotherm models by Pearson method

Variables	Experimental data	Langmuir	Freundlich	Redlich–Peterson
Experimental data	**1**			
Langmuir	0.980	**1**		
Freundlich	0.976	0.965	**1**	
Redlich–Peterson	0.987	0.993	0.989	**1**

Table 5 shows that the *Typha australis* leaves adsorbent is an efficient adsorbent for the removal of ATZ from aquatic environments.

Table 5 q_m values of various adsorbents for ATZ adsorption

Adsorbents	q_m (mg g^{-1})	Supported references
Zeolite-A	0.65	[17]
Moringa oleifera pods	0.629	[19]
Soil	0.022	[25]
Rubber granules	0.47	[26]
Iron nano-composite	0.0117	[27]
Natural sepiolite	0.86	[28]
Typha australis	0.87	This work

3.3 *Effect of Initial Concentration for Column Study*

Fixed-bed columns operations were realized in column with a length of 10 cm. ATZ solutions (5–20 mg/L) were sucked up the column filled with 1.8 g of *Typha australis* at 2 mL min^{-1} via peristaltic pump. The collected ATZ solutions at time intervals are microfiltered, and the concentrations were followed by High Performance Liquid of Chromatography. The maximum column capacity q_{total} (mg) and the adsorbed ATZ herbicide concentration C_{ad} (mg L^{-1}) were determined by the following Eqs. (8) and (9), respectively:

$$q_{tot} = \frac{Q}{1000} \int_{t=0}^{t=t_{tot}} C_{ad} dt \quad (8)$$

$$q_{e(exp)} = \frac{q_{tot}}{W} \quad (9)$$

where

Q: flow rate (mL min^{-1}),

W: total weight of *Typha australis* leaves adsorbent in the column in grams.

The determined breakthrough curves are presented in Fig. 5 whereas the bed capacity values in the continuous column for ATZ removal by *Typha australis* from the resulting aqueous solution are listed in Table 6 (Fig. 6).

Table 6 Bed capacity values for the fixed bed adsorption of ATZ by *Typha australis*

ATZ initial concentration (mg L^{-1})	ATZ solution flow rate (mL mn^{-1})	Bed capacity, q_{bed} (mg g^{-1})
5	2	0.37
10	2	0.58
20	2	1.37

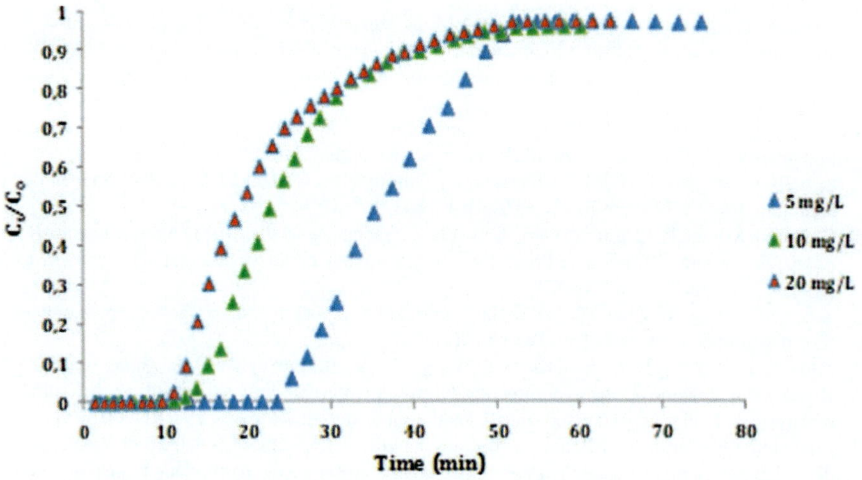

Fig. 6 ATZ adsorption on *Typha australis* (flow rate, 2 mL min^{-1})

As shown in Table 6, if the initial ATZ concentration rises from 5 to 20 mg L^{-1}, the adsorption capacity similarly rises from 0.37 to 1.37 mg g^{-1}. The increase of the adsorption capacity could be due to the increase of the ATZ transfer on the interface and the competition among the ATZ molecules in initial concentration [29].

4 Conclusions

ATZ adsorption was expressed well by the PSO kinetics model. The experimental equilibrium data fit Langmuir perfectly. The Redlich–Peterson isotherm and Pearson method confirmed the suitability of Langmuir. The continuous column experiments have been investigated. These results conclude that native *Typha australis* leaves without any pretreatment could be a promising adsorbent for water treatment. The use of *Typha australis* leaves can be environmentally friendly and may promote green chemistry.

References

1. Chen GC, Shan XQ, Zhou YQ, Shen X, Huang HL, Khan SU (2009) Adsorption kinetics, isotherms and thermodynamics of atrazine on surface oxidized multiwalled carbon nanotubes. J Hazard Mater 169(1–3):912–918
2. Baghapour MA, Nasseri S, Derakhshan Z (2013) Atrazine removal from aqueous solutions using submerged biological aerated filter. J Environ Health Sci Eng 11(1):6

3. Lage ALA, Ribeiro JM, Souza-Fagundes EM, Brugnera MF, Da Silva MDC (2019) Efficient atrazine degradation catalyzed by manganese porphyrins: determination of atrazine degradation products and their toxicity evaluation by human blood cells test models. J Hazard Mater 378:120748
4. Alvarez MD, Fuiman LA (2005) Environmental levels of atrazine and its degradation products impair survival skills and growth of red drum larvae. Aquat Toxicol 74:229–241
5. Sass JB, Colangelo A (2006) European union bans atrazine, while the United States negotiates continued use. Int J Occup Environ Health 12:260–267
6. Gupta R, Kumar P, Fahmi N, Garg B, Dutta S, Sachar S, Matharu AS, Vimaleswaran KS (2020) Endocrine disruption and obesity: a current review on environmental obesogens. Curr Res Green Sustain Chem 3:100009
7. Zhang H, Wang X, Qian M, Jin Y (2022) Atrazine exposure induces hepatic metabolism disorder in male adult zebrafish. Toxics 10(7):400
8. Jiang Z, Li J, Jiang D, Gao Y, Chen Y, Wang W, Cao B, Tao Y, Wang L, Zhang Y (2020) Removal of atrazine by biochar-supported zero-valent iron catalyzed persulfate oxidation: reactivity, radical production and transformation pathway. Environ Res 184:109260
9. Mahlalela LC, Casado C, Marugan J, Septien S, Ndlovu T, Dlamini LN (2020) Photocatalytic degradation of atrazine in aqueous solution using hyperbranched polyethyleneimine templated morphologies of $BiVO_4$ fused with Bi_2O_3. J Environ Chem Eng 8:10425
10. Chan C, Tao S, Dawson R, Wong P (2004) Treatment of atrazine by integrating photocatalytic and biological processes. Environ Pollut 131(1):45–54
11. Ventura A, Jacquet G, Bermond A, Camel V (2002) Electrochemical generation of the Fenton's reagent: application to atrazine degradation. Water Res 36(14):3517–3522
12. Ahmad AL, Tan LS, Shukor SRA (2008) Dimethoate and atrazine retention from aqueous solution by nanofiltration membranes. J Hazard Mater 151(1):71–77
13. Pinto CF, Antonelli R, de Arujo KS, de Toledo ALF, Fernandes DM, Granato AC, Azevedo EB, Malpass GRP (2019) Experimental-design-guided approach for the removal of atrazine by sono-electrochemical-UV-chlorine techniques. Environ Technol 40(4):430–440
14. N'diaye AD, Boudokhane C, Kankou M, Dhaouadi H (2019) Potential of rice husk ash in atrazine removal. Chem Ecol 1–15
15. Adeniyi AG, Igwegbe CA, Ighalo JO (2021) ANN modelling of the adsorption of herbicides and pesticides based on sorbate-sorbent interphase. Chem Africa 4:443–449
16. Boucher J, Steiner L, Marison IW (2007) Bio-sorption of atrazine in the press-cake from oil seeds. Water Res 41(15):3209–3216
17. Jamil TS, Gad-Allah TA, Ibrahim HS, Saleh TS (2011) Adsorption and isothermal models of atrazine by zeolite prepared from Egyptian kaolin. Solid State Sci 13(1):198–203
18. Chaparadza A, Hossenlopp JM (2012) Adsorption kinetics, isotherms and thermodynamics of atrazine removal using a banana peel based sorbent. Water Sci Technol 65(5):940–947
19. Coldebella PF, Klen MRF, Nishi L, Valverde KC, Cavalcanti E, Dos Santos OAA, Bergamasco R (2016) Potential effect of chemical and thermal treatment on the kinetics, equilibrium, and thermodynamic studies for atrazine biosorption by the *Moringa oleifera* pods. Can J Chem Eng 95(5):961–973
20. N'diaye AD, Boudokhane C, Elkory MB, Kankou M, Dhaouadi H (2018) Methyl parathion pesticide removal from aqueous solution using Senegal River *Typha australis*. Water Sci Technol: Water Supply 18(5):1545–1553
21. Guo L, Li G, Liu J, Meng Y, Xing G (2012) Nonlinear analysis of the kinetics and equilibrium for adsorptive removal of Cd (II) by starch phosphate. J Disper Sci Technol 33(3):403–409
22. Salman JM, Njoku VO, Hameed BH (2011) Batch and fixed-bed adsorption of 2,4-dichlorophenoxyacetic acid onto oil palm frond activated carbon. Chem Eng J 174:33–40
23. Aoulad El hadj Ali Y, Demba N'diaye A, Ahrouch M, Sakar EH, Raklami A, Lahcen AA, Stitou M (2022) Dehydrate sewage sludge as an efficient adsorbent for malachite green removal in textile wastewater: experimental and theoretical studies. Chem Africa 1–15
24. Dhaouadi H, M'Henni F (2009) Vat dye sorption onto crude dehydrated sewage sludge. J Hazard Mater 164(2–3):448–458

25. Al-Wabel MI, Abdel-Nasser G, Al-Turki AM, El-Saeid MH (2010) Behavior of atrazine and malathion pesticides in soil: sorption and degradation processes. J Appl Sci 10:1740–1747
26. Alam JB, Dikshit AK, Padhyav B (2000) Efficacy of adsorbents for 2, 4-D and atrazine removal from water environment. Glob Nest Inter J 2:139–148
27. Ali I, ALOthman ZA, Al-Warthan A (2016) Sorption, kinetics and thermodynamics studies of atrazine herbicide removal from water using iron nano-composite material. Int J Environ Sci Technol 13:733–742
28. Gonzalez-Pradas E, Villafranca-Sanchez M, Socias-Viciana M, Fernandez-Perez M, Urena-Amate MD (1999) Preliminary studies in removing atrazine, isoproturon and imidacloprid from water by natural sepiolite. J Chem Technol Biotechnol 74(5):417–422
29. Thuong NT, Nhi NTT, Nhung VTC, Bich HN, Quynh BTP, Bach LG, Nguyen TD (2019) A fixed-bed column study for removal of organic dyes from aqueous solution by pre-treated durian peel waste. Indones J Chem 19(2):486–494

Use of Nanofluid to Improve the Efficiency of Photovoltaic Panel

Badia Rtimi, Ali Benhmidene, Khaoula Hidouri, and Bechir Chaouachi

Abstract Energy demand is consistently expanding due to rapid population growth and economic expansion in the country; however, the reserves of fossil fuels are decreasing. That's why Shift from fossil fuels to renewable energy is recognized worldwide. Solar energy represents the most promising renewable energy source. This source of energy can be converted into electricity using a photovoltaic panel (PV). In hot area, the efficiency of these panel decrease progressively due to the increase in cell temperature. Many researchers use different technology to cool down this panel and enhance its efficiency. The objective of this research is to enhance the performance of the solar panel with the use of a rectangular heat exchanger fixed to the back of PV panels. A comparative study is established between standalone PV panel and PVT panel cooled using water and nanofluid mixture to highlight the utility of using such a system and quantify the amount of gained energy. The use of MgO nanofluid with a mass fraction of 3% improves the electrical power, and the gained energy is equal to 26.28 W/m^2 compared to standalone PV panel.

Keywords PV panel · Cooling · Energy · Nanofluid

B. Rtimi (✉) · A. Benhmidene · K. Hidouri · B. Chaouachi
Laboratory of Energy, Water, Environment and Processes, The National School of Engineering of Gabès, Gabès University, Gabes, Tunisia
e-mail: rtimi.badiaa94@gmail.com

A. Benhmidene
e-mail: Ali.Benhmidene@gmail.com

K. Hidouri
e-mail: khaoula2013@yahoo.fr

B. Chaouachi
e-mail: bechir.chaouachi@enig.rnu.tn

R. Khiari and M. Jawaid (eds.), *Proceedings of the 3rd International Congress of Applied Chemistry & Environment (ICACE-3)*, Springer Proceedings in Materials 23, https://doi.org/10.1007/978-981-99-1968-0_4

1 Introduction

The demand for energy increases proportionally to the growth of the world population, and for other reasons like the use of technology including more recently working from home. According to the International Energy Agency (IEA) report [1], this demand is forecast to rise by 1.3% per year until 2040 while the reserve of fossil energy is decreasing.

Otherwise, fossil fuels will damage our environment and they are associated with high costs. That's why researchers are looking to replace it with a sustainable and efficient energy source. Among renewable energy resources, solar energy is a massive source. In addition it's a clean energy, abundant, and available during the day. Many scientific approaches have been adopted in the aim to increase the production and efficiency of solar energy. The economically efficient solar converters are PV panels and thermal systems. In addition, they are the most adopted in recent years.

The influence of temperature on the efficiency of PV panel, in addition to its low energy conversion present, challenges to the development of PV panel industry [2]. For instance, the performance of crystalline silicon solar cells decreases by 0.5% per an increase of one degree Celsius in solar cell temperature, and this decrease in performance varies depending on the cell type [3]. The heating of the cell leads to a drop in its efficiency and destroys the structure of the solar panel material [4].

Active or passive methods have been developed to reduce the temperature effect [5, 6]. In active systems, the flow is made to circulate using a peripheral power source like a pump or a fan. This technique is the most effective in improving heat removal due to the higher convective heat transfer coefficient. In addition, it is used in research on photovoltaic thermal systems for home heating applications [7]. The main drawbacks of this technique are the high pumping power, the cost of energy, and the incompatibility with tracking [8]. Usually, air, water, and nanofluids are used as cooling fluids. However Water-cooled photovoltaic panels are mostly studied in regards of water thermal conductivity and its availability. The most widely used method of cooling the solar panel is water backflow cooling.

Shalab et al. [9] examined the influence of using a cooling system on the backside PV panel surface. The electrical energy production and efficiency were investigated experimentally. Authors are found an enhancement of 14.1% of the generation power for the proposed cooling system. The Panel PV was benefice of the cooling process. An increase of its electric efficiency from 17.4 to 19.8% was achieved by using the cooling system. In this experimental study by Singth et al. [10], the cooling system is realized with several copper tubes with a diameter of 6.35 mm attached to the backside of the PV panel employing a copper absorber plate. The working medium is water. It was found that a 15.23% reduction in temperature and a 6.08% increase in electrical efficiency were achieved at a mass flow rate of 0.0166 kg/s.

At the beginnings of the twenty-first century and in parallel with technological development, nanomaterials composed of fine ground nanometric particles have been created and introduced in the field of solar energy. These properties relating to thermal storage allow them to be associated with other agents for cooling and thermal storage uses. The CuO–water nanofluid was tested as a HFT by Michael et al. [11] The results showed a 19.25% improvement in the efficiency of the cooled system compared to water.

Tian et al. [12] simulate a simple solar panel by using a cooling system. The indicated cooling system consists of copper tubes disposed in special arrangement and installed at its base, which cools the panel in addition to produce hot water. In this study authors used the MgO/water working fluid. The outcomes show that the efficiency increases by 2.03% for a reduction of flow rate from 0.5 to 4 L/min. In addition, its energy efficiency increases by 0.45% in the case of adding 1% nanoparticles at the working fluid of mass flow of 0.5 L/min.

Jia et al. [13] constructed a PV/T system for testing some nano-particle in water as a coolant in the aim to reduce the temperature cells. According to the experimental results, the PVT efficiency is better in the case of Al_2O_3/water as a working fluid then the collector using TiO_2/water nanofluid. The thermal power of PV/T collector increases by 12.11% by decreasing mass flow rate of nanofluid from 0.03 to 0.0005 kg/s.

In this paper, the overall efficiency of the solar panel is the main studied factor. That's why a rectangular heat exchanger is designed and added to the back of the solar panel to enhance the performance of a standalone PV panel. The mathematical equation that describes the heat transfer mechanism was implemented into Matlab software, and the performance of the PV panel was tested in different cases: without cooling, water-cooling, and nanofluid cooling.

2 The Proposed Photovoltaic System

The proposed PV/T system contains, in addition of, solar cell (PV), glass layer, Tedlar layer, heat exchanger, and an insulating layer as represented in Fig. 1. The role of glass is to transmit solar radiation to the photovoltaic cell. When the PV cell temperature rises, it transfers heat to the Tedlar layer, then to the fluid inside the heat exchanger, and finally a small amount of energy is transferred to the insulating layer. The objective of thermal balance is to benefit from the cooling process for thermal storage and electrical efficiency improvement. In the energy balance, the following assumptions have been made:

- The temperature variation along the system is linear.
- Well developed flow on the driving surface.
- The friction effect is neglected.
- All PVT compounds had same area.
- The adopted meteorological data of the region of Gabes are Latitude = 33° 53′ 17.077″ N, Longitude = 10° 5′ 51.079″ E.
- No dust or shadow on the surface of the cells.

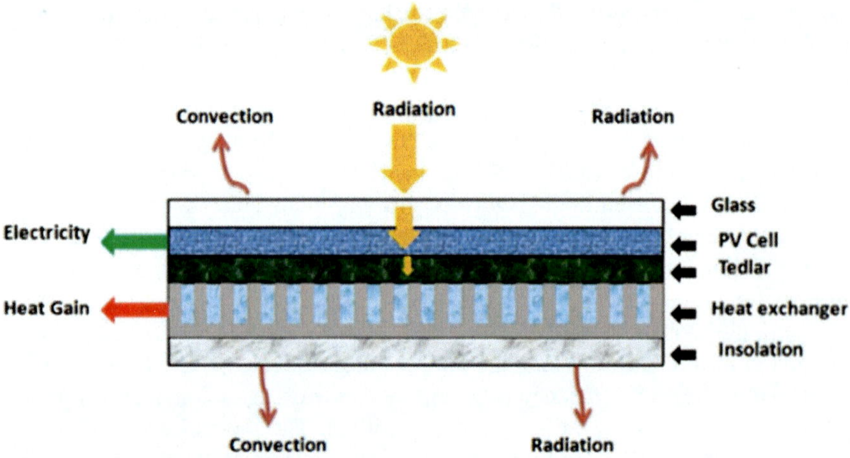

Fig. 1 The PV panel components

3 Modeling

The modeling is obtained using energy balances on the various components of the PV panel.

3.1 Energy Balance on the Glass

The temperature of the glass T_g is given from the energy balance on the glass as follows [14]:

$$\frac{dT_g}{dt} = \frac{1}{m_g c_g}\left(\alpha_g I_G - P_{conv} - P_{ray} - P_c\right) \tag{1}$$

The absorbed power is $P_{in} = \alpha_g I_G$, where αg is the absorption factor of the glass and IG is the solar irradiance. The convective power exchanged with the ambient air is calculated according to equation [14]:

$$P_{conv} = h_{vv} \times A_g \times \left(T_g - T_a\right) \tag{2}$$

where h_{vv} is the heat transfer coefficient by convection due to the wind. It is described by the following equationn [14]:

$$h_{vv} = 5.7 + 3.8 \times U_V \tag{3}$$

The glass exchanges energy by radiation with the sky according to the following equation [14]:

$$P_{ray} = \sigma \times \varepsilon_g \times A_g \times \left(T_g^4 - T_{sky}^4\right) \tag{4}$$

The temperature of the sky is related to the ambient temperature by the following relation [14]:

$$T_{sky} = 0.0552 \times T_a^{1.5} \tag{5}$$

The thermal power exchanged by conduction in cell panel cells P_c is [14]:

$$P_c = \frac{K_g}{\delta_g} \times A_g \times (T_g - T_c) \tag{6}$$

K_g, A_g, and δ_g are the thermal conductivity, the surface, and the thickness of the glass layer.

3.2 *Energy Balance on the PV Cell*

The variation of the cell temperature with time is given by the following differential equation [14]:

$$\frac{dT_c}{dt} = \frac{1}{m_c c_c}\left(\tau_g A_g \alpha_c \beta_c I_G + P_c - P_{ct} - E_c A_c\right) \tag{7}$$

The power exchanged by conduction between the cell and the Tedlar layer is expressed by:

$$P_{ct} = h_{cc} A_c (T_c - T_t) \tag{8}$$

where h_{cc} is the conductive heat transfer coefficient between the cell and the Tedlar layer, it is given by:

$$h_{cc} = \frac{k_c}{\delta_c} + \frac{k_t}{\delta_t} \tag{9}$$

k_c and δ_c are the thermal conductivity and thickness of the PV cell. Similarly, k_t and δ_t are the thermal conductivity and thickness of the Tedlar.

The thermal power absorbed by the cells can be obtained by the equation:

$$P_c = \alpha_c \tau_g \beta_c I_G A_g \eta_c \tag{10}$$

where α_c, β_c, and η_c are the absorption factor, fill factor, and efficiency of the PV cell. τ_g is the transmitivity of the glass. The electrical energy generated by a solar panel depends on the temperature of the cell. T_c, temperature of the referenced cell. $T_{c,\ ref}$, it is calculated as follows:

$$E_c = \alpha_c \tau_g \beta_c I_G A_g \eta_{ref} \left(1 - \theta\left(T_c - T_{c,ref}\right)\right) \tag{11}$$

3.3 *Energy Balance on the Tedlar*

The thermal power stored by the Tedlar layer is determined from the equation below [14]:

$$\frac{dT_t}{dt} = \frac{1}{m_t c_t}(P_{ct} - P_{tf} - P_{rt} + A_t P_t) \tag{12}$$

The power P_{tf} exchanged by convection between the Tedlar and the fluid is given by:

$$P_{tf} = h_{vt} A_t (T_t - T_f) \tag{13}$$

The convective heat transfer coefficient h_{vt} is defined from Nusselt number as follow:

$$h_{vt} = \frac{Nu \times k_f}{D_h} \tag{14}$$

where D_h is the hydraulic diameter.

The power exchanged by radiation between the Tedlar and the pipe walls is a function of the radiative heat transfer h_{rt}, the temperature of the Tedlar T_t, and the temperature of the pipe wall T_{co}. It is given by the following equation:

$$P_{rt} = h_{rt} A_t (T_t - T_{co}) \tag{15}$$

The radiative heat transfer coefficient between the Tedlar layer and the pipe walls is defined as follows:

$$h_{rt} = \frac{\sigma \times (T_t + T_{co})\left(T_t^2 + T_{co}^2\right)}{\frac{1}{\varepsilon_{co}} + \frac{1}{\varepsilon_t} - 1} \tag{16}$$

ε_t and ε_{co} are respectively the emissivities of the Tedlar and the pipe. The thermal power absorbed by the Tedlar (W/m^2) is given by the following expression:

$$P_t = \tau_g(1 - \beta_c)\alpha_t I_G \tag{17}$$

3.4 Energy Balance on the Cooling Fluid

The temperature of the fluid along the panel is deduced by the energy balance given by the following equation [14]:

$$\frac{dT_f}{dt} = \frac{1}{m_f c_f}\left((C_f \dot{m}_f T_{i,f} - C_f \dot{m}_f T_{o,f})A_f + P_{tf} - P_{fco} - P_{ext}\right) \tag{18}$$

$T_{i,f}$ and $T_{o,f}$ are respectively the temperature of the fluid at the inlet and at the outlet. The thermal power exchanged between the fluids and the outside is a function of the Nusselt number, the thermal conductivity of the fluid, the flow section S, and the hydraulic diameter:

$$P_{ext} = Nu \times kf \times S \times \frac{(T_f - T_{amb})}{D_h} \tag{19}$$

The power exchanged by convection between the fluid and the pipe is calculated by [14]:

$$P_{fco} = A_f h_f (T_f - T_{co}) \tag{20}$$

The convective heat transfer coefficient between the fluid and the pipe is given by [14]:

$$h_f = \frac{Nu \times k_f}{D_H} \tag{21}$$

The Nusselt number is a function of the Reynolds and Prandtl numbers. The thermal efficiency of the fluid is calculated using the following expression [14]:

$$\eta_{th} = \frac{C_f \dot{m}_f T_{o,f} - C_f \dot{m}_f T_{i,f}}{S \times I_G} \tag{22}$$

3.5 Nanofluid Properties

- Heat specific capacity [15]:

$$cp_{nf} = (1 - \varphi)cp_{water} + \varphi cp_{np} \tag{23}$$

- Density [16]:

$$\rho_{\text{nf}} = (1 - \varphi)\rho_{\text{water}} + \varphi\rho_{\text{np}} \tag{24}$$

- Dynamic viscosity [17]:

$$\mu_{\text{nf}} = \left(1 + 2.5\varphi + 6.2\varphi^2\right)\mu_{\text{water}} \tag{25}$$

- Thermal conductivity [18]:

$$k_{\text{nf}} = k_{\text{water}} \times \left(\frac{k_{\text{np}} + 2 \times k_{\text{bf}} + 2 \times \varphi\left(k_{\text{np}} - k_{\text{bf}}\right)}{k_{\text{np}} + 2 \times k_{\text{bf}} - 2\varphi\left(k_{\text{np}} - k_{\text{bf}}\right)}\right) \tag{26}$$

4 Comparative Study

To highlight the impact of the use of cooling fluid on the performance of PV panels, a comparative study is conducted in this section. Indeed, the cell temperature, as well as the electrical efficiency, are simulated in three cases; the first one is in absence of cooling medium, the second one is water cooling with a mass flow rate of 0.01 kg/s, and the third one is the use of MgO nanofluid with 3% mass fraction and a flow rate of 0.01 kg/s.

The simulation results for cell temperature and electrical efficiency are shown in Figs. 2 and 3 respectively.

The cell temperature of the stand-alone PV panel, shown in Fig. 2, increases with solar radiation, and the maximum-recorded value is 111.86 °C. The addition of a heat exchanger at the back of the PV model decreases the temperature significantly.

The solar radiation falling on the cells increases its temperature under the influence of the greenhouse effect caused by the glass layer. This heat would be absorbed by the fluid circulating in the heat exchanger pipe. The effect of water-cooling on the electrical efficiency of the conventional PV panel and the PVT system is shown in Fig. 3 where the mass flow rate of water is 0.01 kg/s. The study shows that the efficiency of the conventional PV panel is lower than that of the PVT system.

Based on the results shown in Fig. 2, the addition of a cooling system allows us to reduce the cell temperature from 111.86 to 49.13 °C, representing a reduction of 62.73 °C. This significant temperature reduction improves electrical efficiency. This significant temperature reduction improves the electrical efficiency (Fig. 3). In the case of maximum sunlight, the electrical efficiency decreases to 9.14% in the case of a standalone PV panel, but in the PVT system, this value increases to 13.37%.

In terms of electrical power as it is shown in Table 1, the PV system produces only 50.76 W/m^2 while the PVT system using water as a cooling medium produces

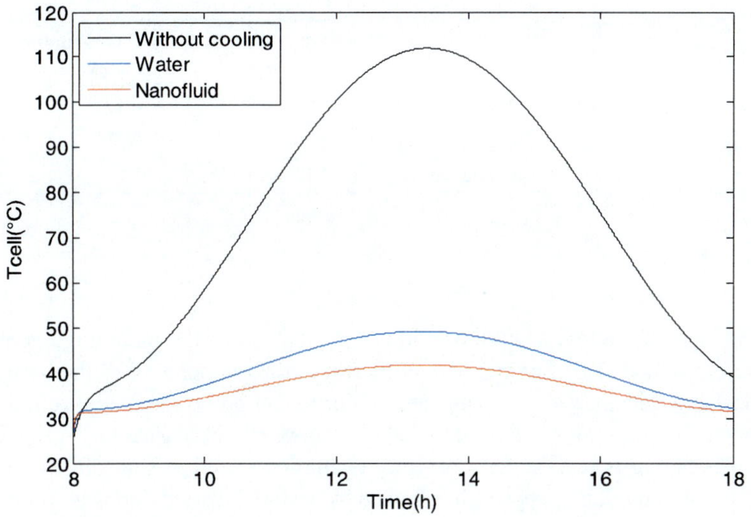

Fig. 2 Cell temperature in different cases

Fig. 3 Electrical efficiency profile in different cases

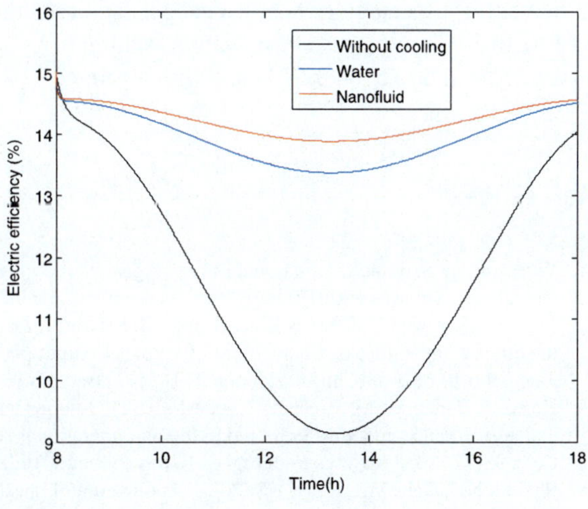

64.61 W/m^2, so a gain of 13.85 W/m^2 is recorded. This gain could increase to 22.81 W/m^2 using the MgO nanofluid in the same conditions.

Another advantage for cooling the photovoltaic panels is the recuperation of thermal energy from the heated water or nanofluid. The gain recorded in the case of using water is about 72.93 W m^{-2}, but this gain can reach 466.09 W m^{-2} by adding a small quantity of nanofluid.

Table 1 Maximum energy produced by PVT system

	Maximum electrical energy (W m^{-2})	Maximum thermal energy (W m^{-2})
MgO-nanofluid	73.57	466.09
Water	64.61	72.93
Without cooling	50.76	0

5 Conclusion

In the process of photovoltaic conversion, a percentage of the incoming energy is converted into heat. This thermal energy increases the temperature of the solar cell and leads to a decrease in performance. To further improve the overall efficiency of solar panel a cooling system was used to remove the excessive heat generated. Nanofluid cooling is one of the promising technology used in solar fields. In this paper, a thermal modeling approach followed by a Matlab simulation was conducted to quantify the amount of heat flow between the different components of the solar panel and the heat exchanger.

In fact, the MgO nanofluid with a 3% of mass fraction and 0.01 kg/s of water circulated in a rectangular heat exchanger increases the electrical efficiency from 9.14 to 13.37% in the case of maximum sunlight. As a result, the electrical power increases from 50.76 to 73.57 W/m^2 with a gain of 22.81 W/m^2.

References

1. Worldenergy outlook, International Energy Agency, www.iea.org
2. Choudhary, Srivastava (2019) https://www.un.org/sustainabledevelopment/energy/ 2020
3. Amelia AR, Irwan YM, Leow WZ, Irwanto M, Safwati I, Zhafarina M (2016) Investigation of the effect temperature on photovoltaic (PV) panel output performance. Int J Adv Sci Eng Inf Technol 6(5):682–688. https://doi.org/10.18517/ijaseit.6.5.938
4. Kurnik J, Jankovec M, Brecl K, Topic M (2011) Outdoor testing of PV module temperature and performance under different mounting and operational conditions. Sol Energy Mater Sol Cells 95(1):373–376. https://doi.org/10.1016/j.solmat.2010.04.022
5. Sharma NK, Gaur MK, Malvi CS (2021) Application of phase change materials for cooling of solar photovoltaic panels: a review. Mater Today: Proc. https://doi.org/10.1016/j.matpr
6. Siah Chehreh Ghadikolaei S (2021) Solar photovoltaic cells performance improvement by cooling technology: an overall review. Int J Hydrogen Energy 46(18):10939–10972. https://doi.org/10.1016/j.ijhydene.2020.12.164
7. Terashima K, Sato H, Ikaga T (2020) Development of an environmentally friendly PV/T solar panel. Sol Energy 199:510–520. https://doi.org/10.1016/j.solener.2020.02.051
8. Tashtoush B, Ballah A, Algharbawi R (2019) Parametric study of a novel hybrid solar variable geometry ejector cooling with organic rankine cycles. Energy Convers Manag 198:111910. https://doi.org/10.1016/j.enconman.2019.111910
9. Shalaby SM, Elfakharany MK, Moharram BM, Abosheiasha HF (2022) Experimental study on the performance of PV with water cooling. Energy Rep 8:957–961. https://doi.org/10.1016/j.egyr.2021.11.155

10. Singh K, Singh D, Kandpal DC, Kumar R (2021) Experimental performance study of photovoltaic solar panel with and without water circulation. Mater Today: Proc 46:6822–6827. https://doi.org/10.1016/j.matpr.2021.04.393
11. Michael JJ, Iniyan S (2015) Performance analysis of a copper sheet laminated photovoltaic thermal collector using copper oxide–water nanofluid. Sol Energy 119:439–451. https://doi.org/10.1016/j.solener.2015.06.028
12. Tian MW, Khetib Y, Yan SR, Rawa M, Sharifpur M, Cheraghian G, Melaibari AA (2021) Energy, exergy and economics study of a solar/thermal panel cooled by nanofluid. Case Stud Therm Eng 28:101481. https://doi.org/10.1016/j.csite.2021.101481
13. Jia Y, Ran F, Zhu C, Fang G (2020) Numerical analysis of photovoltaic-thermal collector using nanofluid as a coolant. Sol Energy 196:625–636. https://doi.org/10.1016/j.solener.2019.12.069
14. Rtimi R, Benhmidene A, Chaouachi B (2022) Effect of nanofluid flow in a rectangular heat exchanger on the performance of a photovoltaic panel. J Therm Sci Eng Appl 1–32. https://doi.org/10.1115/1.4054500
15. Raud R, Hosterman B, Diana A, Steinberg TA, Will G (2017) Experimental study of the interactivity, specific heat, and latent heat of fusion of water based nanofluids. Appl Therm Eng 117:164–168. https://doi.org/10.1016/j.applthermaleng.2017.02.033
16. NabatiShoghl S, Jamali J, Keshavarz Moraveji M (2016) Electrical conductivity, viscosity, and density of different nanofluids: an experimental study. Exp Therm Fluid Sci 74:339–346. https://doi.org/10.1016/j.expthermflusci.2016.01.004
17. Sarsam WS, Amiri A, Zubir MNM, Yarmand H, Kazi SN, Badarudin A (2016) Stability and thermophysical properties of water-based nanofluids containing triethanolamine-treated graphene nanoplatelets with different specific surface areas. Colloids Surf, A 500:17–31. https://doi.org/10.1016/j.colsurfa.2016.04.016
18. Said Z, Saidur R, Hepbasli A, Rahim HA (2014) New thermophysical properties of water based TiO_2 nanofluid—the hysteresis phenomenon revisited. Int Commun Heat Mass Transfer 58:85–95. https://doi.org/10.1016/j.icheatmasstransfer.2014.08.034

Kinetic Study of Dyeing (PET/PTT) Bicomponent Polyester Textiles Using Ecofriendly Carriers

Marwa Souissi, Ramzi Khiari, Mohamed Abdelwaheb, Mounir Zaag, Nizar Meksi, and Hatem Dhaouadi

Abstract In this paper, three ecofriendly carriers are used as biobased adjuvants to dye (polyethylene terephthalate (PET)/polytrimethylene terephthalate (PTT)) bicomponent filaments at 100 °C without degrading their characteristic mechanical properties, more precisely their excellent elasticity. These biobased carriers present an excellent remedy to toxic/chemical carriers and allow savings in terms of dyeing time and dyeing temperature. This paper introduces a kinetic study of dyeing polyesters filaments with three different disperse dyes at 100 °C (by introducing ecofriendly carriers) and at 130 °C (without introducing carriers). Four Kinetics models were studied to explain the adequate dyeing mechanism of studied filaments with disperse dyes after introducing ecofriendly carriers.

1 Introduction

Nowadays, bicomponent filaments are increasingly used in the clothing sector (sportswear, swimsuits, bathtub, etc.). These filaments are crimped due to the two parallel bicomponent filaments (PET/PTT) attached with different shrinkage or expansion properties and the internal molecular rearrangement of the fibrous material [1–4].

M. Souissi (✉) · M. Abdelwaheb · N. Meksi · H. Dhaouadi
Laboratory of Environmental Chemistry and Cleaner Process (LCE2P-LR21ES04), University of Monastir, 5019 Monastir, Tunisia
e-mail: souissi.marwa20@yahoo.com

M. Souissi · N. Meksi
National Engineering School of Monastir, University of Monastir, 5019 Monastir, Tunisia

R. Khiari
Higher Institute of Technological Studies (ISET) of Ksar-Hellal, 5070 Ksar-Hellal, Tunisia

CNRS, Grenoble INP, LGP2, University of Grenoble Alpes, F-38000 Grenoble, France

M. Zaag
Société Industrielle des Textiles (SITEX), 5070 Ksar-Hellal, Tunisia

R. Khiari and M. Jawaid (eds.), *Proceedings of the 3rd International Congress of Applied Chemistry & Environment (ICACE-3)*, Springer Proceedings in Materials 23, https://doi.org/10.1007/978-981-99-1968-0_5

The use of bicomponent filaments (PET/PTT) allows to develop textile materials with the following performances: a low percentage of elongation and shrinkage which allows fabrics subsequently manufactured to retain their shape [1–4], exceptional comfort [5, 6], good elasticity and elastic recovery in all directions [7, 8], a softer, smoother feel than textured yarns, and an excellent resistance to chemical substances [9, 10].

Due to their high crystallinity, the diffusion rate of disperse dyes inside the fiber is relatively low at the usual dyeing temperatures, around 100 °C. Special dyeing processes must therefore be used. So the first process consists in dyeing at 100 °C in the presence of certain carriers and the second consists of dyeing under pressure at high temperature [11, 12]. However, the dyeing of polyester fibers in full bath at high temperature is certainly simpler and requires fewer chemicals [10], but this method is very consuming in terms of energy.

Indeed, in order to increase the dyeing speed and to dye at low temperature, it is recommended to introduce dyeing adjuvants called carriers. These agents can be classified into four groups: phenolic compounds, primary amines, hydrocarbons, and esters [13, 14]. They are present in the dye bath in the form of an emulsion [10].

Pasquet et al. studied the toxicity of carriers used in the dyeing of polyester filaments with disperse dyes: benzoic acid, p-dichlorobenzene, o-dichlorobenzene, diphenyl, and phenylphenol. Obtained results showed that phenylphenol is the most toxic and carcinogenic carrier followed by p-dichlorobenzene which is slightly toxic and possibly carcinogenic [17]. All carriers present huge dangers because of their toxicity, bad odor, and volatile ability. Therefore, the dyeing method with these carriers is far from being the best solution for dyeing PET fibers with disperse dyes [15–17].

Several works were interested in the study of the kinetics of dyeing polyesters filaments and concluded that the dyeing mechanism of PET fibers by disperse dyes can be summarized in three steps. The first step involves dissolving part of the dye in the dye bath. Then, the second step is the adsorption of the dye dissolved in the dye bath on the surface of the PET fiber. Finally, it is the diffusion of the dye inside the PET fiber. The rate of dye adsorption by different polyester fibers (PET, PTT, PBT, etc.) is affected by different factors, including the fineness of the fiber, the bath ratio, and the diffusion coefficient of the dye as well in the aqueous medium than in the fiber. The diffusion of dyes in the fibers is linked to the molecular order, to the crystal lattice and to the morphology of the polyester filaments. Therefore, in the case of PET and PTT dyeing, for example, the diffusion coefficient is not the same [10, 13, 16].

The most important parameters which influence the diffusion coefficient are the affinity of the dye, the swelling of the fiber, the enthalpy change required for formation of free volume, and glass transition temperature. Various experiments were carried out to establish a correlation between the diffusion coefficients and fiber structure and two main models have been developed to explain the mechanisms of migration inside polymers: the "pore" model and the "free volume" model [10, 13, 16].

In the literature, the study of the Kinetics dyeing of bicomponent filaments has not been the subject of several studies. Then, it is essential to find an adequate

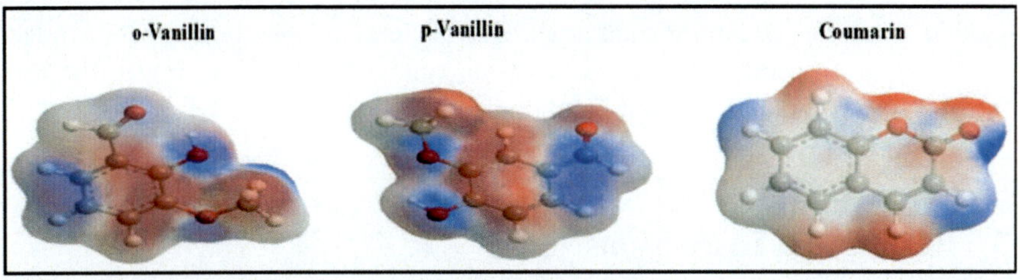

Fig. 1 Chemical configurations of biobased carriers

dyeing mechanism to describe the dyeing of bicomponent filaments composed of two filaments having behaviors that can be different during dyeing. This paper studies the kinetics of (PET/PTT) bicomponent filament in the two cases: dyeing at low temperature equal to 100 °C with introducing of biobased carriers and dyeing at high temperature equal to 130 °C without adding carriers. Based on four models of kinetics, the adequate dyeing mechanism was established.

2 Material and Methods

2.1 Reagents

In this study, three different disperse dyes were used in the dyeing of bicomponent filaments: CI Disperse Red 167.1, CI Disperse Yellow 211, and CI Disperse Red 60 having three different molecular weights: high, medium, and low, respectively. The effect of the addition of three biobased carriers, namely: o-vanillin, p-vanillin and coumarin, on the dyeing kinetics at 100 °C of bicomponent filaments was investigated. Figure 1 shows the chemical configurations of these carriers.

2.2 Textile Support

A 100% bicomponent filament knit made with bicomponent filaments (60% polyethylene terephthalate/40% polytrimethylene terephthalate) was produced using a circular knitting machine with a diameter of 12 cm.

2.3 The Kinetics and Isotherms Models Used

The study of the dyeing kinetics of bicomponent filaments with disperse dyes without or with adding ecological carriers was carried out based on four models namely:

pseudo first order, pseudo second order, Elovich, and intra-particle models. Indeed, the study of adsorption isotherms has been established by referring to the three theories: Nernst, Langmuir, and Freundlich models.

3 Results and Discussions

3.1 Adsorption Kinetics Results

According to the nonlinear kinetics equations, instantaneous adsorbed quantity was estimated and presented according to the variation of dyeing time. Figure 2 shows the dyeing kinetics of bicomponent filaments (60% PET, 40% PTT) by disperse dyes at 100 °C using respectively 6 g L^{-1} of o-vanillin. From Fig. 2, it can be seen that the pseudo first order model presents the best results in term of correlation coefficients. Obtained results for the other biobased carriers, p-vanillin and coumarin, are similar to those obtained for o-vanillin.

3.2 Adsorption Isotherms Results

The study of dye isotherms was carried out for each of the studied dyes (with low, medium, and high energy) at concentrations equal to 0.05 g L^{-1} up to 4 g L^{-1}. The dyeing is carried out under pressure at 130 °C. Figure 3 shows the different dyeing isotherms. Obtained results make to deduce that the Nernst linear model would be the adequate model to describe the adsorption of studied dyes into (PET/PTT) polyesters filaments.

3.3 Dyeing Mechanism

The dyeing mechanism can be summarized by the following steps (Fig. 4). So at the beginning we have an interaction between the molecules of dye particles in the water and the carrier. Carriers are largely absorbed by polyesters fibres. They dislocate and push macromolecular chains of the PET fiber and freeing up spaces. So, the dye can penetrate more quickly inside the fiber [13–16].

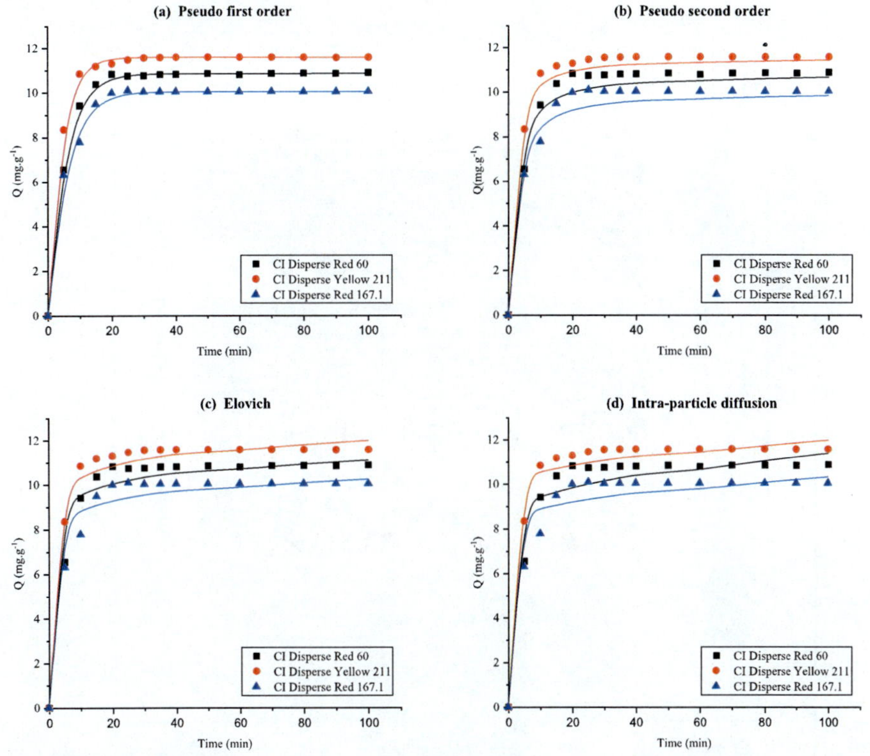

Fig. 2 Adsorption kinetics models of dyeing (PET/PTT) polyesters filaments after introducing 6 g L^{-1} of o-vanillin

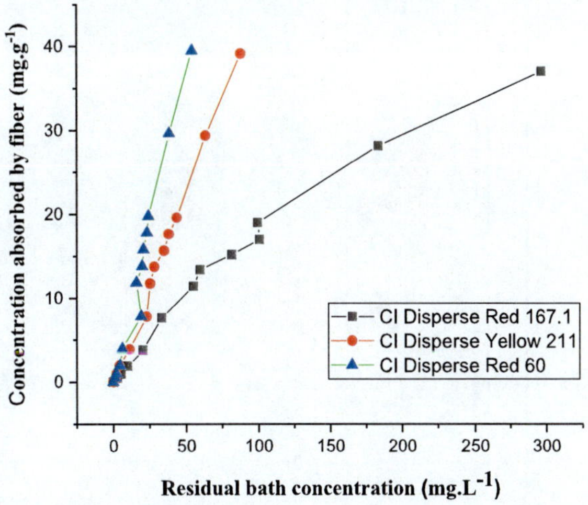

Fig. 3 Dyeing isotherms of (PET/PTT) polyesters filaments

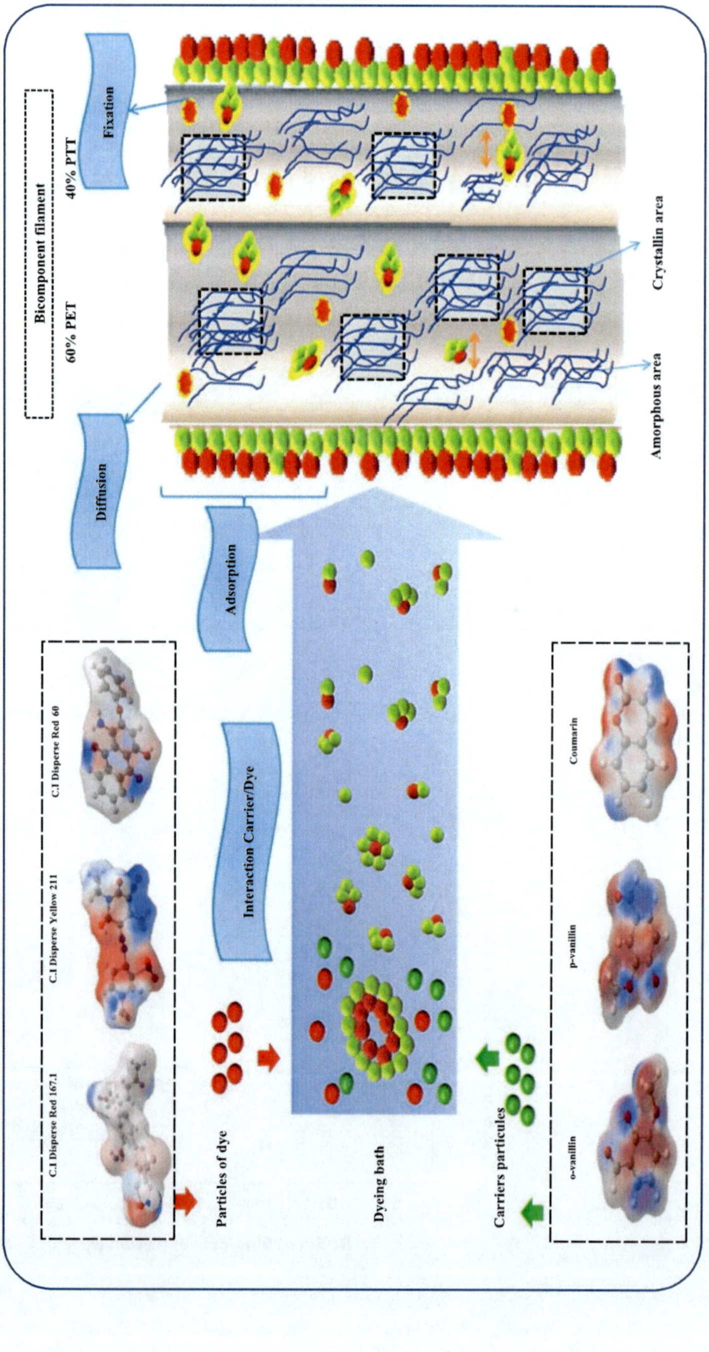

Fig. 4 Dyeing mechanism of (PET/PTT) bicomponent filament using ecofriendly carriers

4 Conclusion

In this paper, studied (PET/PTT) polyesters filaments have undergone dyeing using three classes of disperse dyes (high, medium, and low) at dyeing temperatures of 130 °C and at 100 °C with introducing three different biobased carriers. The two dyeing processes presented excellent dyeing yields. The environmentally friendly process which introduces carriers has saved energy and dyeing time. The dyeing kinetics were established and studied. The pseudo first order kinetics model was the appropriate model to present the adsorption of studied dyes into bicomponent filament in the case of the two different dyeing processes. It was proved also that the dyeing isotherm follows the Nernst model.

References

1. Souissi M, Khiari R, Zaag M, Meksi N, Dhaouadi H (2020) Effect of the morphology of polyesters filaments on their physical properties and dyeing performances. Polym Bull 78:2685–2707
2. Souissi M, Khiari R, Haddar W, Zaag M, Meksi N, Dhaouadi H (2020) Dyeing of innovative bicomponent filament fabrics (PET/PTT) by disperse dyestuffs: characterization and optimization process. Processes 8:501
3. Souissi M, Khiari R, Zaag M, Meksi N, Dhaouadi H (2022) Bicomponent polyesters filaments with high performances: study of their stability before and after dyeing with low, medium, and high energy disperse dyes. Fibers Polym 23:2828–2837
4. Souissi M, Khiari R, Zaag M, Meksi N, Dhaouadi H (2022) Comparative study of dyeing performances of advanced polyesters filaments. Springer Proc Mater 17:261–267
5. Jeong HM, Ahn BK, Cho SM, Kim BK (2000) Water vapor permeability of shape memory polyurethane with amorphous reversible phase. J Polym Sci Pol Phys 38:3009–3017
6. Ozdil N, Anand S (2014) Recent developments in textile materials and products used for activewear and sportswear. Elect J Veh Tech 8:68–83
7. Bansal P, Maity S, Sinha SK (2018) Elastic recovery and performance of Denim fabric prepared by cotton/Lycra Core Spun Yarns. J Nat Fibers 8:1184–1198
8. Jin L, Fumei W, Bugao X (2010) Factors affecting crimp configuration of PTT/PET bi-component filaments. Text Res J 81:538–544
9. Land MC, Bruck HAA (2002) Fundamental investigation into large strain recovery of one-way shape memory alloy wires embedded in flexible polyurethanes. Smart Mater Struct 11:130–139
10. Broadbent AD (2001) Basic principles of textile coloration. Society of Dyers and Colourists, Sherbrooke
11. Tavanaie MA, Shoushtari AM, Goharpey F (2010) Polypropylene/poly(butylene terephthalate) melt spun alloy fibres dyeable with carrier-free exhaust dyeing as an environmentally friendlier process. J Clean Prod 18:1866–1871
12. Carrion Fité FJ (1995) Dyeing polyester at low temperatures: kinetics of dyeing with disperse dyes. Text Res J 65:362–368
13. Burkinshaw SM (1995) Chemical principles of synthetic fibre dyeing. Blackie Academic & Professional, London
14. Iskender MA, Becerir B, Koruyucu A (2005) Carrier dyeing of different energy level disperse dyes on polyester fabric. Tex Res J 75:462–465
15. Souissi M, Khiari R, Zaag M, Meksi N, Dhaouadi H (2021) Ecological and cleaner process for dyeing bicomponent polyester filaments (PET/PTT) using ecological carriers: analysis of dyeing performance. RSC Adv 11:25830–25840

16. Souissi M, Khiari R, Zaag M, Meksi N, Dhaouadi H (2022) Kinetic study of dyeing bicomponent polyester textiles (PET/PTT) using environmentally friendly carriers. RSC Adv 12:2361–2374
17. Pasquet V, Perwuelz A, Behary N, Isaad J (2013) A potential carrier for low temperature dyeing of polyester fabrics. J Clean Prod 43:20–26

Preliminary Physicochemical and Phytochemical Study of Seeds of *Ziziphus mauritiana*

Mohamed Moutaly, Ould El Moustapha Abdallahi, and Abdoulaye Demba N'diaye

Abstract In this study, the nutritional composition of the seeds of the *Ziziphus mauritiana* were analyzed using standard procedures. The percentage composition of the *Ziziphus mauritiana* seeds is moisture 5.1%, ash 7.2%, protein 11.56%, and lipids 8.45%. *Ziziphus mauritiana* seeds are rich in protein and lipids which are of potential industrial significance. Calcium, potassium, and magnesium constitute the major minerals of *Ziziphus mauritiana* seeds. This fact is of great economic interest owing to several applications of *Ziziphus mauritiana* seeds in the food, especially in the third world. The presence of several phytochemical compounds showed high therapeutic potential of *Ziziphus mauritiana,* and it can take for medicinal purposes after determining the seeds pharmacologically.

Keywords *Ziziphus mauritiana* · Seeds · Nutritional · Proximate · Mineral

1 Introduction

Jujube trees belong to the genus Ziziphus Miller (1764), Zizyphus Jussieu (1789), to the family Rhamnaceae and to the order Rhamnales. They include between 86 and 170 species. The fruit is the main interest of the jujube tree, widely consumed by the populations and the object of a lucrative trade involving mainly women.

M. Moutaly
Laboratoire de Chimie, Institut National de Recherches en Santé Publique, BP 695 Nouakchott, Mauritanie

O. E. M. Abdallahi
Laboratoire de Biotechnologie Alimentaire, Département des Sciences et Technologies des Aliments, Institut Supérieur d'Enseignement Technologique, Rosso, Mauritanie

A. D. N'diaye (✉)
UR Eau Pollution et Environnement, Département de Chimie, Faculté des Sciences et Techniques, Université de Nouakchott, BP 880 Nouakchott, Mauritanie
e-mail: abdouldemba@yahoo.fr

R. Khiari and M. Jawaid (eds.), *Proceedings of the 3rd International Congress of Applied Chemistry & Environment (ICACE-3)*, Springer Proceedings in Materials 23, https://doi.org/10.1007/978-981-99-1968-0_6

The jujube tree appears to be one of the spontaneous fruit trees that best withstand high temperatures and long dry periods (sometimes from 6 to 12 months) [11]. Its ecological plasticity makes it more interesting especially for Sahelian countries in a context of desertification and decline in agricultural production. In addition, the species is widely used in agroforestry systems such as defensive hedges [12]. The jujube tree plays an important role in human nutrition.

Jujube trees are shrubs that include many species, the best known of which in the southern Sahara is *Ziziphus mauritiana* Lam. variety orthacantha (DC) [8].

The fruit of the jujube tree is a glabrous, globose, or oval drupe that varies greatly in diameter (1.2–1.5 cm). Some cultivated oval varieties reach up to 3–5 cm in diameter. The fruits contain a large stone wrapped in a whitish pulp more or less floury. The skin is green, turning yellow-green and then reddish-brown as they ripen [18]. They behave more like climacteric fruits. They are of better quality when picked at the right stage of ripeness. Indeed, immature fruits are not sweet and have an astringent taste, while the organoleptic characteristics of overripe fruits the organoleptic characteristics of overripe fruit decrease considerably.

Ziziphus mauritiana fruits are nutritious and are usually consumed fresh or dried [20, 24]. The fruit has high sugar content which provides energy and also has protein with many essential amino acids [24].

The nutritional properties of jujube pulp are widely recognized. Its richness in vitamin C and A, anti-oxidant molecules, and minerals such as phosphorus and calcium makes it a fruit of great nutritional value. The chemical composition varies considerably with the variety and also with the stage of maturity.

The use which is made of the various parts of the jujube tree is multiple. Indeed, the leaves are used as fodder for cattle, goats, and sheep. The leaves, roots, and bark are used for medicinal purposes. However, the fruit is the main interest of the jujube tree. It is consumed in various forms by local populations and is the subject of intense trade, which mainly involves women. They are consumed fresh, in the form of juice, or are processed into flour for various food uses: dough, cakes, drinks, porridge [4, 23]. When they are processed into flour, they are dried in the sun for 8–10 days and then pounded in a mortar to separate them from the pits without the latter breaking and to process them into flour in the same operation [7, 15].

The seeds extract of *Ziziphus mauritiana* showed a hypoglycaemic effect on alloxan-induced diabetic mice [7] and improved lipid composition in the serum of dietary hyperlipidemic animals [21]. The seed also can be used to treat diarrhea, nausea, and vomiting during pregnancy [14].

Ziziphus mauritiana seeds are a potential plant both for food and phytomedicine applications. Therefore, this study aimed to evaluate the nutritional composition of *Ziziphus mauritiana*. The present study could help in assessing the actual nutrient composition of these seeds and also, help to supplement the nutrients needs and demand of man and other domestic animals. In this study, the proximate analysis such as moisture, ash, protein ad lipid, and the minerals such as calcium, magnesium, sodium, potassium, iron, and zinc were used. Qualitative phytochemical screening of *Ziziphus mauritiana* seed such as flavonoids, steroids, terpenoids, anthraquinones, and Phlobatannins was carried out.

Fig. 1. *Ziziphus mauritiana* **a**: fruits; **b**: seeds; **c**: seed powder

2 Material and Methods

2.1 Sample Collection and Preparation

All the chemicals and reagent used were of analytical grade. The Zizyphus mauritiana fruits were purchased by a vendor from Sebkha market, Nouakchott City in Mauritania. The seeds were washed by distilled water and subjected to shade dry during five days. The seeds were crushed in powder and stored in dessicator before use (Fig. 1).

2.2 Proximate Composition Analysis

The moisture was determined as described by Karaye et al. [19]. The proteins and ash were determined using the method described by AOAC [2]. The lipids were determined using the method described by Folch et al. [13].

2.3 Determination of Mineral Contents in Ziziphus mauritiana Seeds

Calcium, magnesium, sodium, potassium, iron, and zinc were evaluated by placing 0.5 g of the *Ziziphus mauritiana* seed powder samples in a Teflon digestion tube. 7 mL of 65% concentrated HNO_3 and 1 mL of 30% H_2O_2 was also added. The tubes are then introduced into the microwave mineralizer (Aton Paar, France) until complete mineralization. The mineralization solutions are transferred into 50 mL flasks and then filled up with distilled water. Calcium, magnesium, zinc, and iron were determined by the absorption spectrometric (Spectrometer Atomic Absorption PGG 990) method using the standard calibration curve of Calcium, magnesium, zinc, and iron, respectively. The sodium and potassium contents were analyzed by the emission photometric (Flame Photometric FP 640) method using the standard calibration curve of sodium and potassium, respectively.

2.4 *Qualitative Phytochemical Screening of Zizyphus mauritiana Seeds*

Qualitative phytochemical screening of Zizyphus mauritiana seeds such as flavonoids, steroids, terpenoids, anthraquinones, and Phlobatannins was carried out of the Zizyphus mauritiana seeds samples using standard procedures described by several reports [25].

3 Results and Discussion

3.1 *Ziziphus mauritiana Seeds Proximate Composition*

Proximate component of the seeds of Ziziphus Mauritania are presented in Table 1. Ziziphus Mauritania seeds showed low moisture content (5.01%). This value is lower than values obtained for Ziziphus Mauritania seeds (29.42%) [16, 17]. The ash content of Ziziphus Mauritania seeds is 7.2%. The values obtained for ash content (7.2%) for seed were higher than values reported for *Ziziphus mauritiana* seeds (2.79%) [32] and Ziziphus Lotus seeds (1.05%) [6]. Ash content gives an idea about the minerals content of the food material [5]. This implied that the seed was rich in mineral contents. The protein content of Ziziphus Mauritania seeds is 11.56%. The values obtained for protein (11.56%) for seed were higher than values reported for *Ziziphus mauritiana* seeds (6.64%) [17]. However, the values obtained for protein content for seed were lower than values reported for Ziziphus lotus seeds (19.11%) [9]. The high protein content indicates that the seed could be an important source for protein supplement. Lipid content of Ziziphus Mauritania seeds is 8.45% which is higher than the 1.89% reported by Jailani et al. [17].

Table 1 *Ziziphus mauritiana* seeds proximate compositions

Parameters	Mean
Moisture (%)	5.01
Ash (%)	7.2
Protein (%)	11.56
Lipids (%)	8.45

3.2 Mineral Contents of *Ziziphus mauritiana* Seeds

The mineral salts contained in the seeds of *Ziziphus mauritiana* are summarized in Table 2. The main minerals are Calcium, Magnesium, Potassium, Sodium, Iron, and Zinc.

There are more than twenty mineral salts. They are essential elements involved in many mechanisms of the human body. They contribute to the proper functioning of the body. Mineral salts are generally supplied by food or certain medications. They are classified into two categories: major minerals and trace elements. The major minerals (Ca, Mg, P, Na, K, and Cl) have a structural and functional role. As for the trace elements, the most essential for life or health, that is to say, those whose insufficient intake leads to functional disturbance and that supplementation with this element prevents or corrects this disturbance are Co, Cr, Cu, Fe, I, Mo, Se, Zn, Mn, Sn, As, Ni, Si, and V. Trace elements are in the milligram or microgram level per kilogram of body weight while major minerals are in the gram per kilogram range and require intakes of more than 100 mg per day.

Of all the minerals determined, showed that calcium is the most abundant having values of 170.62 mg/100 g. It is followed by potassium (40 mg/100 g), sodium (30 mg/100 g), and magnesium (25.32 mg/100 g). Indeed, Calcium is an intracellular second messenger in the insulin hormone response and its electro-physiological potentialities make it an ion of great importance [28]. It improves glucose tolerance. In case of deficiency, its metabolism may be directly involved in the development of non-insulin dependent diabetes. Its deficiency not only establishes diabetes but may also be responsible for the degenerative complications characteristic of type 2 diabetes [16, 31]. Sodium avoids the loss of liquid from the body and plays a major role a major role in water balance and cell permeability. According to Oudot et al. [29], excessive consumption leads to insulin resistance. Low sodium supplemented with potassium attenuates insulin resistance. Na/K ratio in the body is important because it helps in controlling high blood pressure. Na/K ratio of less than one is recommended [26]. The Na/K ratio obtained in this study is about 0.75, which indicate that the seeds of *Ziziphus mauritiana* could be useful in the lowering blood pressure. Magnesium is an enzyme activator. It participates in the major metabolisms in the body. It raises insulin secretion, and facilitates glucose utilization [22]. Deficiency is a cause of Insulin Resistance. Deficiency of this mineral has been associated with diabetes and

Table 2 Minerals composition of *Ziziphus mauritiana* seeds (mg/100)

Mineral elements	Mean
Ca	170.62
Mg	25.32
K	40
Na	5
Fe	4.5
Zn	3.21

Table 3 Phytochemical constituent of the seeds of the *Ziziphus mauritiana*

Phytochemical	Observations
Flavonoids	–
Steroids	+
Terpenoids	+
Anthraquinones	–
Phlobatannins	–

a high risk of retinopathy in diabetics [3]. The adrenal glands play an essential role in regulating sodium retention and excretion. Studies have also shown that Mg will affect adrenal cortical activity and results in increased in Mg retention [3]. Iron content of the seeds (8.5 mg/100 g) is higher than the 1.21% reported by Yerima and Adamu [32]. Zinc content of the seeds Ziziphus Mauritania (3.52 mg/100 g) is higher than the 1.89% reported by Yerima and Adamu [32]. Iron is an important element in the diet of pregnant women, nursing mothers, and infants [27], while zinc could play a major role in normal body development. Indeed, zinc, a coenzyme, is the support of a good activity because its deficiency is the cause of insulin resistance [1].

3.3 Qualitative Phytochemical Analysis

The phytochemical analysis of the *Ziziphus mauritiana* seeds are presented in Table 3. The results of the phytochemical screening test revealed the presence of steroids, and terpenoids, while flavonoids, saponins and anthraquinones were absent. According to researchers [10, 30], the presence of these phytochemicals showed high therapeutic potential for the seeds of the *Ziziphus mauritiana* and it can take for medicinal purposes.

4 Conclusions

The proximate and minerals analysis showed that the seeds from *Ziziphus mauritiana* fruits could be considered as rich in protein, lipids, and minerals elements. The result suggests that the seeds of *Ziziphus mauritiana* fruits could contribute greatly toward meeting human nutritional requirement for normal growth and adequate protection against disease arising from malnutrition. Presence of several phytochemical compounds such as steroids and terpenoids showed high therapeutic potential of *Ziziphus mauritiana,* and it can take for medicinal purposes after determining the seeds pharmacologically.

References

1. Abitogun AS, Borokini FB, Ashogbon AO (2010) Nutritional and chemical composition of ripe and unripe Vitex grandifolia. J Res Natl Dev 8(1):1–8
2. AOAC (1995) Official methods of analysis, 16th edn. Association of Analytical Chemists, Washington, DC. Afr J Food Sci 4:115–119
3. Aremu MO, Olaofe O, Akintayo ET (2006) Compositional evaluation of cowpea (Vigna unguiculata) varieties and scarlet runner bean (Phaseolus coccineus) varieties flour. J Food Agric Environ 4(2):39–43
4. Azam-Ali S, Bonkoungou E, Bowe C, DeKok C, Godara A, Williams JT, Ber and Other Jujubes (2006) Williams JT, Smith RW, Haq N, Dunsigaer Z (eds) Fruits for the future. Southampton Centre for Underutilized Crops, Southhampton, p 1, 19 and 29
5. Bello MO, Falade OS, Adewusi SR, Olawore NO (2008) Studies on the chemical composition and anti-nutrients of some lesser known Nigeria fruits. Afr J Biotechnol 7(21):3972–3979
6. Berkani F, Serralheiro ML, Dahmoune F, Mahdjoub M, Kadri N, Dairi S, Achat S, Remini H, Abbou A, Adel K, Madani K (2021) Ziziphus lotus (L.) Lam. plant treatment by ultrasounds and microwaves to improve antioxidants yield and quality: an overview; Nor. Afr J Food Nutr Res 5(12):53–68
7. Bhatia A, Mishra T (2010) Hypoglycemic activity of *Ziziphus mauritiana* aqueous ethanol seed extract in alloxan induced diabetic mice. Pharm Biol 48:604–610
8. Chevalier A (1952) Les jujubiers ou Ziziphus du Sahara [in French]. Revue Internationale de Botanique Appliquée et D'agriculture Tropicale (Paris) 32:574–577
9. Chouaibi M, Mahfoudhi N, Rezig L, Donsì F, Ferrari G, Hamdi S (2012) Nutritional composition of Zizyphus lotus L. seeds. J Sci Food Agric 92(6):1171–1177
10. Chukwuebuka E, Chinenye IJ (2015) Biological functions and anti-nutritional effects of phytochemicals in living system. J Pharm Biol Sci 10(2):10–19
11. Depommier D (1988) *Ziziphus mauritiana* Lam., culture et utilisation en pays Kapsiki (NordCameroun). Bois For Trop 218:57–63
12. Djimdé M (1997) *Ziziphus mauritiana* a live hedge species. Atelier pan Africain sur *Ziziphus mauritiana*. ICRISATIICRAF Bamako, p 18
13. Folch J, Lees M, Sloane Stanley GH (1957) Une méthode simple pour l'isolement et purification des lipides totaux des tissus animaux. J Bouillir Chem 226(1):497–509
14. Goyal M, Sasmal D, Nagori B (2012) Review on ethnomedicinal uses, pharmacological activity and phytochemical constituents of *Ziziphus mauritiana* (Z. jujuba Lam., non Mill). J Complement Med Drug Discov 2(2):101–116
15. Gupta MK, Bhandari AK, Singh RK (2012) Pharmacognostical evaluation of the leaves of *Ziziphus mauritiana*. Int J Pharm Sci Res 3(3):818–821
16. Hassan LG, Umar KJ, Sa'idu I, Folorunsho FA (2008) Proximate, mineral and anti nutritional factors of Daniella oliveri seeds kernel. Chemclass J 5:31–36
17. Jailani FNAM, Zaidan UH, Rahim MBHA, Gani SSA, Halmi MIE (2020) Evaluation of constituents and physicochemical properties of Malaysian underutilized *Ziziphus mauritiana* (Bidara) for nutraceutical potential. Int J Fruit Sci 20(3):394–402
18. Kaaria S (1998) *Ziziphus mauritiana*—a valuable tree for arid and semi-arid lands. Fact sheet. A quick guide to multipurpose trees from around the world. Morrilton, Arkansas: a publication of the Forest, Farm and Community Tree Network (FACT Net), Winrock International, p 4
19. Karaye IU, Ladan MU, Adili IS, Shehu A, Lawal HM, Sahabi MH (2020) Phytochemistry and proximate composition of fruit pulp and seed of desert date, *Balanites aegyptiaca* (Del.). Int J Sci Glob Sustain 6:2
20. Khoshoo TN, Subrahmanyam GV (1985) Ecodevelopment of arid lands in India with nonplants for arid lands. In: Wickens GE, Goodin JR, Field DV (eds) Plants for arid lands. Unwin Hyman, London, pp 243–265
21. Kim HS (2002) Effects of the *Zizyphus jujuba* seed extract on the lipid components in hyperlipidemic rats. Int J Food Sci Nutr 7(1):72–77

22. Lee SH, Jouihan HA, Cooksey RC, Jones D, Kim HJ, Winge DR, McClain DA (2013) Manganese supplementation protect against diet-induced diabetes in wild type mice by enhancing insulin secretion. Endocrinology 154(3):1029–1038
23. Marwat SK, Khan MA, Khan MA, Rehman FU, Ahmad M, Zafar M, Sultana S (2009) *Salvadora persica, Tamarix aphylla* and *Zizyphus mauritiana*-three woody plant species mentioned in Holy Quran and Ahadith and their ethnobotanical uses in north western part (D.I. Khan) of Pakistan. Pak J Nutr 8:542–547
24. Memon AA, Memon N, Luthria DL, Pitafi AA, Bhanger MI (2012) Phenolic compounds and seed oil composition of *Ziziphus mauritiana* L. Fruit. Pol J Food Nutr Sci 62(1):15–21
25. Obadoni BO, Ochuko PO (2002) Phytochemical studies and comparative efficacy of the crude extracts of some haemostatic plants in Edo and Delta States of Nigeria. Glob J Pure Appl Sci 8(2):203–208
26. Ogunlade I, Olaofe O, Fadare T (2005) Chemical composition, amino acid and functional properties of Leucaena leucocepha seeds flour. Niger J Appl Sci 21:7–12
27. Oluyemi EA, Akinlua AA, Adebayo MB (2006) Mineral contents of some commonly consumed Nigerian foods. Sci Focus 11(1):153–157
28. Ouazar MA, Amine M, Harifi G, Ouilki I, Younsi R, Belkhou A, Bouchti IE, Hassani SE (2010) Evaluation of the calcium intake in population of Marrakesh and its region: 1000 cases. Ann Endocrinol 71:51–55
29. Oudot C, Lajoix AD, Jover B, Rugale C (2012) Oxidative stress and beneficial effect of sodium restriction on kidney damage associated with insulin resistance in rats. Annales de Cardiologie et d'Angéiologie 61(3):162–166
30. Patel SS, Savjani JK (2015) Systematic review of plant steroids as potential anti-inflammatory agents: current status and future perspectives. J Phytopharm 4(2):121–125
31. Vexiau P, Cathelineau G, Luyckx A, Lefebvre P (1986) Amelioration of glucose tolerance and correction of reactive hypoglycemias induced by intravenous calcium infusion cannot be explained by modifications in blood glucagon levels. Diabete Metab 12(4):177–185
32. Yerima BI, Adamu HM (2011) Proximate chemical analysis of nutritive contents of Jujube (*Ziziphus mauritiana*) seeds. Int J Phys Sci 6(36):8079–8082

Characterization of Natural Tow Fibers

Elaissi Arwa, Ghith Adel, and Alibi Hamza

Abstract The natural waste from two fibers has been gathered during the gypsum manufacturing sectors and recycled by fraying after cleaning fibers to reduce the detrimental effects on the environment. A thermogravimetric analysis (ATG) exhibits that the thermal degradation of fibers starts from 350 °C. The Differential scanning calorimeter analysis (DSC) curves show a peak at 85–90 °C which can be attributed to water loss. These cellulosic fibers have been blended with cotton fibers to make them non-woven and improve the performance of the material. The findings demonstrate that variations in the proportions of (cotton/tow) fibers have an effect on mechanical strength (Tear test, traction.). The chemical modification reduces resistance to water penetration chemical modification reduces resistance to water penetration. Roughness assessments of non-wovens using a User Surface Tester (UST) reveal that an increase in the amount of tow compared to cotton causes an increase in the roughness of samples. These fibers can be used in several fields such as construction, automotive, and also as composite reinforcements.

Keywords Fibers · Tow · Waste · Characterization · Natural

1 Introduction

Nowadays, manufacturing processes are developed with a more conscious consideration towards the protection of the environment and the planet. This fact is also manifested by an increase in research developing the use of industrial wastes. Studies have demonstrated that using natural fibers for automotive composite applications,

E. Arwa (✉) · G. Adel
Textile Materials and Research Process (MPTEX), Tunisia National Engineering School of Monastir (ENIM), University of Monastir, Monastir, Tunisia
e-mail: elaissiarwa@gmail.com

A. Hamza
Laboratory of Thermal and Energetic Studies (LESTE), Tunisia National Engineering School of Monastir, University of Monastir, Monastir, Tunisia

R. Khiari and M. Jawaid (eds.), *Proceedings of the 3rd International Congress of Applied Chemistry & Environment (ICACE-3)*, Springer Proceedings in Materials 23, https://doi.org/10.1007/978-981-99-1968-0_7

such as kenaf, ramie, flax, hemp, and cotton, has several benefits on both a technical and financial level [1]. Productivity and flexibility are two of these technologies' most significant benefits (can produce various products for various applications fields). The work by Tufail Hassan et al. focuses on the analysis of the mechanical, thermal, and acoustic properties of green epoxy composites reinforced using used natural fibers [2]. However, in a composite, the matrix, usually a polymer, is hydrophobic, whereas natural fibers are hydrophilic which creates fiber/matrix interface problems. Chemical treatments on the fibers, such as scouring and cationization, are anticipated to address this issue. In this context, investigate the printing and impact properties of a polyester/cotton composite made of cationized hollow cellulosic fiber [3]. Using quaternary ammonium compounds, Kim and Choi have also researched and analyzed the cationization of periodate-oxidized to increase the cellulose's affinity [4].

2 Material and Methods

2.1 *Materials Preparation*

To eliminate contaminants, desizing is applied to all fibers first. They are treated in a solution of 1.5 mL/L of NaOH, 3 g/L of Na_2CO_3, and 4 g/L of nonionic detergent for 120 min. To stop the emulsion from degrading and contaminants from precipitating on the surface, fibers were then rinsed in hot and cold water, and then allowed to air dry [5]. The cationization of fibers was studied with a [polyimethyldiallylammonium-diallylamine chloride] (PDDACD) copolymer, under alkaline conditions To create a homogeneous fiber web, fibers are opened, cleaned, and blended using a card machine in accordance with a mass percentage of (tow/cotton). The sail is strengthened by the needling process. The manufacture of nonwovens is achieved in the ISET-KH (High Institute of Technological Studies in Ksar-Hellal, Tunisia).

2.2 *Thermogravimetric Analysis*

The thermal stability of composite materials was investigated using the TGA method in a nitrogen environment at a rate of 10 °C/min from 25 to 700 °C. The phase transition temperature and latent heat of form-stable composites were measured at a rate of 10 °C/min under a nitrogen atmosphere, ranging from −25 to 300 °C, using a DSC instrument coupled with TGA equipment. The curves were captured using a Simultaneous DSC-TGA apparatus from Perkin Elmer that was placed in the GEMTEX facility in France.

2.3 Roughness Analysis

Utilizing the universal surface tester, roughness is measured (UST). It establishes the micromechanical characteristics of materials at the surface and was created by the company Innowep in partnership with the University of Applied Sciences in Darmstadt. This kind of examination offers pertinent details on the overall deformation, elastic deformation, and permanent deformation of the material under the study's surface. The purpose of the probe, a pressure transducer (also known as a pressure sensor), is to transform electrical voltage into an amplitude value. The papillary, 1.8 mm-diameter stylus that was chosen is comparable in size to the 3D printer's extrusion head. The speed of the software is 0.1 mm/s. In a length of 10 mm, it scans every 0.5 mm. Therefore, the measurement range is ±2000 μm [6].

According to Fig. 1, R_{ZI} roughness values were determined for each sample.

According to the formula [7], the average roughness R_z value for each sample was determined:

$$R_Z = \frac{1}{n_l} \times \sum_{0}^{l} R_{ZI} \tag{1}$$

where

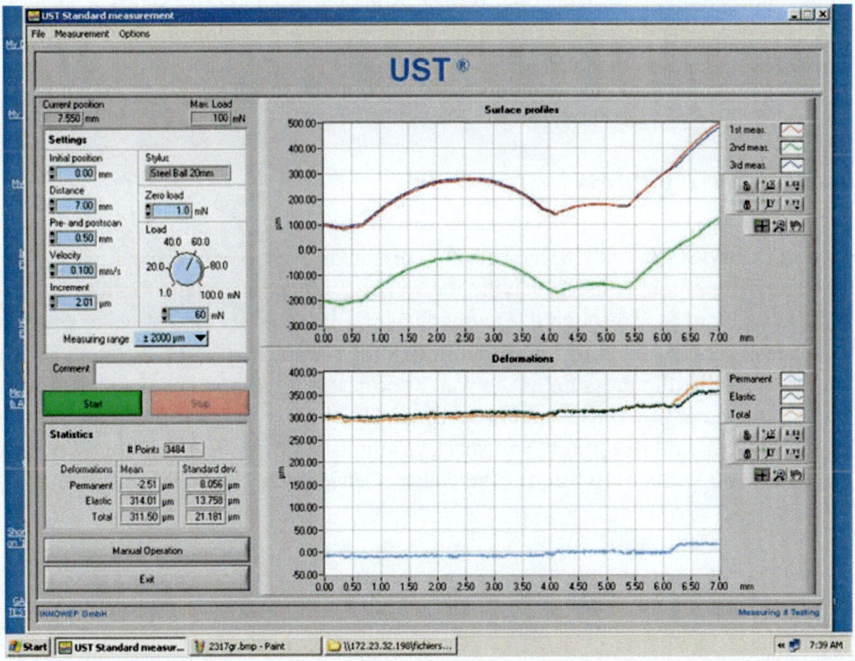

Fig. 1 Example of roughness calculation

R_{Zl}: the predicted maximum heights for each base length;

l: base length;

n_l: number of repetitions.

2.4 Tensile Tests

A short bar of the material under study is clamped between the jaws of a traction machine for this test, which pulls until the bar snaps. In the framework, EN ISO 13937 has been prepared.

A tensile test is a physics experiment that enables an axial stress state to assess the elastic behavior of a material and quantify the material's degree of breaking strength.

2.5 Tear Test

By ISO 9073, the machine's speed was set at 100 mm/min, and the jaws were spaced apart by (25 ± 1) mm. Nonwoven samples should be cut to the typical template's dimensions of (75 ± 1) mm × (150 ± 2) mm for an isosceles trapezoid. The trapezoid's non-parallel edges must be fixed. It is then applied to the specimen such that the tear covers the full width after a continuous increase in elongation. The average maximum tear strength is calculated as a result.

2.6 Water Tightness Test

According to ISO 811: 2018, hydrostatic pressure is used to determine the resistance to water penetration. A fabric's hydrostatic charge serves as a gauge for how difficult it is for water to pass through it. On one face of a 100 cm^2 specimen, water pressure is gradually increased until it passes through the specimen at three different points. At the third site, the pressure as the water permeates the fabric is noted. The rate of increase in water pressure is said to be 10 cm H_2O/min ± 0.5 cm H_2O/min or 60 cm H_2O/min ± 3 cm H_2O/min (98 mbar H_2O/min ± 0.5 mbar H_2O/min or 58.5 mbar H_2O/min ± 2.9 mbar H_2O/min). The water is leave in the test tube at 20 ± 2 °C.

Table 1 The nonwovens roughness values

Samples	Fibers pre-treatment	% of fibers	R_Z (μm)
1	Raw	20/80	1000
2	Raw	50/50	1100
3	Raw	80/20	1300
4	After scouring	20/80	950
5	After scouring	50/50	1100
6	After scouring	80/20	1150
7	Cationized 3%	20/80	800
8	Cationized 3%	50/50	1050
9	Cationized 3%	80/20	1100
10	Cationized 6%	20/80	1000
11	Cationized 6%	50/50	1050
12	Cationized 6%	80/20	1100
13	Bleached	20/80	950
14	Bleached	50/50	100
15	Bleached	80/20	1100

3 Results and Discussions

3.1 Roughness Evaluation Test

According to Table 1, nonwovens get rougher as the percentage of the yarn compared to cotton increases. The tow fiber's roughness explains this. Thus, the roughness values are decreased by moving from the raw sample to samples that have undergone chemical modification.

The measured deformations for a sample's surface width of 10 mm are shown in Fig. 2. As can be observed, although having equal levels of roughness, untreated samples 1 and 2 exhibit severe levels of the overall strain. This fact explains how chemical alteration, which involves the ablation of surface impurities, affects the supports. The treatment effect is smoother and nearly linear once surface pollutants and other loosely bound species have been removed [8].

3.2 Water Impermeability

Chemical alteration lowers the resistance to water penetration, as seen in Table 2. Comparing cationized non-wovens to untreated non-wovens, impermeability values slightly decreased [6].

With increased roughness, fiber water permeability decreases. By adding percentages of tow that have the roughness aspect, for instance from an untreated sample

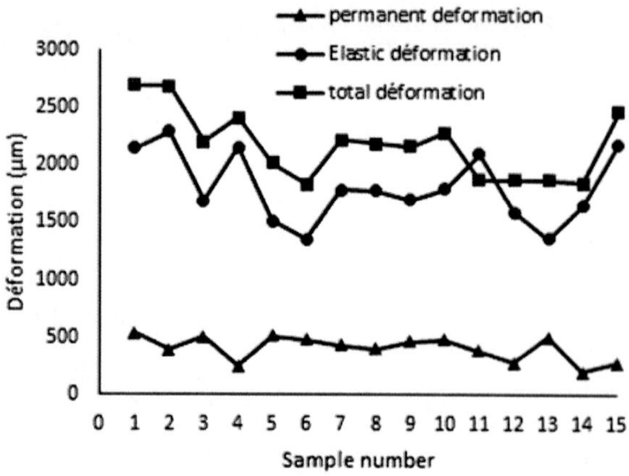

Fig. 2 The estimated parameters of the 3D surface texture for analyzed samples

Table 2 The values of resistance to water penetration for nonwovens

Samples	Water impermeability (mbar/min)
1	15.8
2	16.4
3	17.1
4	14.9
5	15.9
6	16.23
7	12.1
8	13.3
9	14.02
10	14.6
11	14.8
12	10.2
13	11.9
14	12.4
15	12.65

of 20/80 to an untreated sample of 50/50, the roughness value increases, and as a result, the permeability to water decreases. This kind of behavior is consistent with the research findings of Akhavan et al. [7] in the literature.

Table 3 Results of the nonwovens' tensile test

Samples max force	Maximal force (N)	Deformation at break (%)	Elongation at max force (mm)
1	8.701	79.586	55.008
2	6.506	32.503	44.808
3	8.700	79.587	55.008
4	5.2	43.683	78.909
5	5.836	40.238	57.229
6	6.73	28.754	25.475
7	4.22	36.57	45.66
8	3.08	35.283	65.253
9	3.22	32.6	43.75
10	7.24	47.090	70.663
11	7.6	47.359	81.542
12	7.625	34.44	51.943
13	7.2	33.88	47.82
14	7.65	34.56	46.55
15	7.32	33.68	46.87

3.3 Tensile Test at Break

Table 3 shows that the tensile strength is marginally decreased by the chemical treatment. The hornification process could be the root of the decrease in mechanical resistance. This effect is related to cellulosic products' porosity, which is caused by a reduction in the pores' accessibility to water in the case of cationization [9, 10].

4 Conclusion

Waste cellulosic fibers were acquired, cleaned, transformed, and mixed with cotton fibers for the development of nonwovens in this study. The impact of the cationization of fibers on some properties is studied. The products obtained can be applied in several fields such as automotive, aeronautics (reinforcements of composites), and also in the field of craftsmanship and furnishings.

References

1. Mueller, Krobjilowski A, Muessig J (2001) Molded fiber-reinforced parts-special properties and influencing parameters. In: Proceedings of the 2001 beltwide cotton, conferences, Anaheim, CA
2. Hassan T, et al (2020) Acoustic, mechanical and thermal properties of green composites reinforced with natural fibers waste. J Polym 12(3):654
3. Hakeim O, et al (2017) Printing and impact properties of cationised hollow cellulosic fiber based polyester/cotton composite. J Carbohydr Polym 171:59–67
4. Kim JY, Choi H-M (2014) Cationization of periodate-oxidized cotton cellulose with choline chloride. Cellul Chem Technol 48(1–2):25–32
5. Jabli M, et al (2011) Adsorption of acid dyes from aqueous solution on a chitosan-cotton composite material prepared by a new pad-dry process. J Eng Fibers Fabr 6(3)
6. Patanaik A, Anandjiwala R (2009) Some studies on water permeability of nonwoven fabrics. Text Res J 79(2):147–153
7. Akhavan A, Rajabipour F (2012) Quantifying the effects of crack width, tortuosity, and roughness on water permeability of cracked mortars. Cem Concr Res 42(2):313–320
8. Vrinceanu N, et al (2018) Correlation between surface engineering and deformation response of some natural polymer fibrous systems. J Eng Fibers Fabr 13(2)
9. Han W, et al (2012) Study on the modification of bleached eucalyptus kraft pulp using birch xylan. Carbohydr Polym 88(2):719–725
10. Hu G, et al (2015) Adsorption of cationized eucalyptus heteropolysaccharides onto chemical and mechanical pulp fibers. Carbohydr Polym 123:324–330

Effect of the Entomopathogenic Fungus *Lecanicillium Muscarium* on the Whitefly *Bemisia Tabaci* in a Cucumber Greenhouse

Besma Hamrouni Assadi, Sabrine Chouikhi, Ali Benhmidene, Naima Anayed, and Mohamed Sadok Belkadhi

Abstract Heated greenhouses are a favorable environment for the development of pests. To reduce the excessive use of chemical pesticides against *Bemisia tabaci* in greenhouses in southern Tunisia, the improvement of biological control methods is necessary such as entomopathogenic fungi EPF. Microbiological control through the use of EPF is one of the means of control that preserves the environment. The pathogenicity of a commercial formulation based on *Lecanicillium muscarium* was tested on eggs and young and old larvae of *B. tabaci* in a geothermal cucumber greenhouse. The effect of *L. muscarium* was compared to two chemical pesticides based on Acetamiprid and Thiamethoxam and an untreated control. This fungus caused high reduction rates of eggs and larvae comparable to those caused by chemical pesticides. In the last week of monitoring, this fungus causes a reduction in eggs, second-stage larvae and fourth-stage larvae of *B. tabaci* respectively of 81.81%, 84.52% and 67.63%. As a result, *L. muscarium* can be used against *B. tabaci* in a geothermal greenhouse.

Keywords *Bemisia tabaci* · Environment · Geothermal greenhouse · *Lecanicillium muscarium*

B. H. Assadi (✉) · S. Chouikhi · M. S. Belkadhi
Dryland and Oases Cropping Laboratory, Arid Regions Institute, Street El Jorf, 4119 Medenine, Tunisia
e-mail: bessadi_27@yahoo.fr

B. H. Assadi · N. Anayed · M. S. Belkadhi
Technical Center for Protected and Geothermal Crops (CTCPG Gabes Tunisia), Gabès, Tunisia

A. Benhmidene
High Institute of Applied Sciences and Technology of Gabes, Gabès, Tunisia

N. Anayed
Laboratory of Energy, Water, Environment and Processes, The National School of Engineering of Gabès, Gabès University, Gabès, Tunisia

R. Khiari and M. Jawaid (eds.), *Proceedings of the 3rd International Congress of Applied Chemistry & Environment (ICACE-3)*, Springer Proceedings in Materials 23, https://doi.org/10.1007/978-981-99-1968-0_8

1 Introduction

The areas of protected and geothermal crops in Tunisia have followed a spectacular evolution in recent years. This sector faces several phytosanitary constraints, of which the whitefly *Bemisia tabaci* is one of the most harmful pests. This pest causes direct and indirect damage causing huge yield losses [1, 2]. By feeding on the phloem, *B. tabaci* secretes large amounts of honeydew which reduces the rate of photosynthesis. This whitefly transmits many plant viruses [3, 4]. Protected and geothermal crops in Tunisia are offered to this whitefly. Faced with the abusive use of chemical pesticides, this pest is developing resistance. The development of biological control methods in greenhouses is becoming a necessity, in particular the use of EPF. These microorganisms are recognized for their effectiveness against insect pests and cause reductions in the population of *B. tabaci* [5–7]. *L. muscarium* is a widely occurring entomopathogenic fungus [8].

2 Materials and Methods

2.1 Bioessays

The tests took place in a cucumber greenhouse (*Cucumis sativus*) with a dimension of 9 × 60 m. A biological treatment based on *L. muscarium* strain 1979: 48 g/kg ($1.1*10^{10}$ spores/g) was compared with two chemical pesticides based on Acetamiprid and Thiamethoxam and an untreated control against *B. tabaci*. During the trial period, no other treatment was used in the greenhouse. The tests were carried out according to the Randomized Complete Blocks Design. Each crop line designates a block which is then divided into four elementary plots of 20 m. Each elementary plot is treated by one of the treatments according to the diagrams illustrated in Fig. 1.

The pathogenicity of *L. muscarium* has been evaluated against the eggs, young larvae and old larvae of *B. tabaci*. The bioassays take place during the period from April 3 to May 5, 2020. The products used are T1: *L. muscarium*; T2: Acetamiprid; T3: Thiamethoxam; T4: untreated control.

Weekly sampling was carried out one month before and after treatment. For each elementary plot, twelve leaves were randomly sampled from four plants. Each sheet was set in an artificial sack, accompanied by a label. The samples collected are kept at 4 °C until observation under a binocular magnifying glass. The efficacy of the products was evaluated by calculation of the reduction rates of eggs, and second- and fourth-stage larvae of *B. tabaci*. Reduction of *B. tabaci* by the treatments was calculated according to the Henderson and Tilton formula (Henderson Tilton as flow):

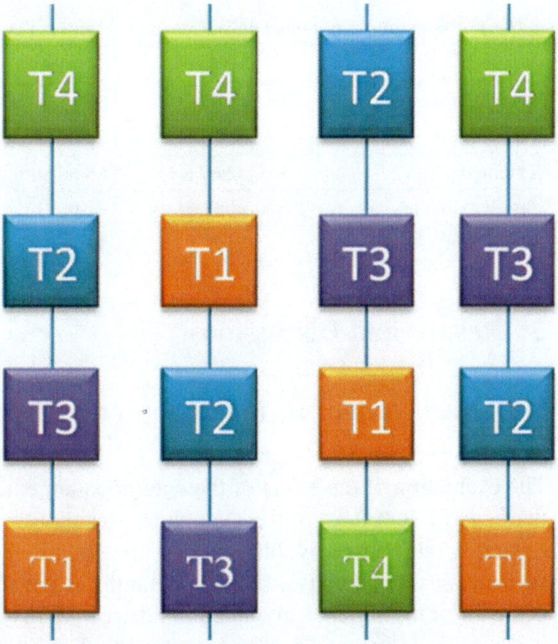

Fig. 1 Diagram of the experimental treatment protocol

$$\text{Reduction}\ (\%) = 1 - \left(\frac{\begin{pmatrix} N_0 \text{ in control} \\ \text{before treatement} \end{pmatrix} \times \begin{pmatrix} N_0 \text{ in treatement} \\ \text{after treatement} \end{pmatrix}}{\begin{pmatrix} N_0 \text{ in control} \\ \text{after treatement} \end{pmatrix} \times \begin{pmatrix} N_0 \text{ in treatement} \\ \text{before traitement} \end{pmatrix}} \right) \times 100$$

2.2 Statistical Analyses

For variance analysis (ANOVA), the reduction egg and larval populations of the whitefly were subjected to unidirectional analysis associated with a Tukey HSD test ($p < 0.05$) using the XLSTAT 2019 software, Microsoft Excel.

Table 1 Reduction of *B. tabaci* eggs

	10 April	17 April	24 March	05 May
Lecanicillium muscarium	75.17 a	80.79 a	82.46 a	80.71 a
Acetamiprid	81.83 a	84.09 a	82.49 a	85.42 a
Thiamethoxam	82.67 a	85.66 a	82.72 a	81.81 a
Pr > F (Modèle)	<0.0001	<0.0001	<0.0001	<0.0001
Significant	Yes	Yes	Yes	Yes

3 Results and Discussions

3.1 Reduction of B. Tabaci Eggs Caused by Treatments

The evaluation of the effect of the entomopathogenic fungus *L. muscarium* against the eggs of *B. tabaci* in the cucumber greenhouse is shown in Table 1. The egg reduction rate caused by this entomopathogenic fungus was compared to that of the two chemical treatments based on Acetamiprid and Thiamethoxam for a month. The behavior of the variance in reduction rates caused by the three treatments shows that it is weakly influenced by these treatments. Indeed, the reduction rates of the eggs entrained by the three treatments are comparable. Biological treatment based on *L. muscarium* has a high pathogenicity on *B. tabaci* eggs over time causing mortality rates exceeding 75.17%.

3.2 Reduction of the Second Instar of B. Tabaci

The effect of the two chemical treatments and the entomopathogenic fungus *L. muscarium* on the second instar larvae of *B. tabaci* is shown in Table 2. Statistical analyses show that the reduction rates of the second instar larvae caused by the three treatments are comparable. Moreover, the L2 larval reduction rates are greater than 78.22% along the test. Similarly, the rate of reduction of second instar larvae of *B. tabaci* caused by the two chemical treatments is high over time, exceeding 79.79%.

3.3 Reduction of the Fourth Instar of B. Tabaci

The reduction of fourth instar larvae of *B. tabaci* caused by *L. muscarium* in the cucumber greenhouse is compared with those of the chemical treatments based on Acetamiprid and Thiamethoxam (Table 3). Statistical analyses show that the three treatments have a comparable insecticidal effect during the first two weeks. From

Table 2 Reduction of the second instar of *B. tabaci*

	10 April	17 April	24 March	05 May
Lecanicillium muscarium	79.49 a	78.22 a	82.15 a	83.94 a
Acetamiprid	81.74 a	79.79 a	89.95 a	82.69 a
Thiamethoxam	83.26 a	84.16 a	82.95 a	84.52 a
Pr > F (Modèle)	<0.0001	<0.0001	<0.0001	<0.0001
Significant	Yes	Yes	Yes	Yes

Table 3 Reduction of the fourth instar of *B. tabaci*

	10 April	17 April	24 March	05 May
Lecanicillium muscarium	65.06 a	66.02 a	57.15 b	51.08 b
Acetamiprid	67.88 a	64.98 a	76.57 a	64.77 a
Thiamethoxam	68.23 a	65.37 a	67.87 ab	67.63 a
Pr > F (Modèle)	<0.0001	<0.0001	<0.0001	<0.0001
Significant	Yes	Yes	Yes	Yes

the third week of treatment, the two chemical treatments are more effective than the biological treatment based on *L. muscarium.* Indeed, the reduction rates of L4 larvae caused by *L. muscarium* did not exceed 66.02% over time.

It turns out, according to these tests, the entomopathogenic fungus *L. muscarium* showed a significant pathogenic power on each stage of the growth of *B. tabaci* tested. Both eggs and second-stage larvae are the most susceptible to the entomopathogenic fungus in the cucumber greenhouse. While the fourth instar of *B. tabaci* is the most resistant, this agrees with previous studies testing the same fungus against *B. tabaci* in geothermal greenhouses in southern Tunisia on tomato and pepper crops. This fungus shows significant efficacy on eggs and adults of *B. tabaci* for tomato and pepper crops [9]. In addition, the utilization of two entomopathogenic fungi *L. muscarium* and *Beauveria bassiana* under controlled conditions has shown significant efficacy versus the different steps of the growth of *B. tabaci* [10]. The control of insect pests has been improved by the use of entomopathogenic fungi [11]. The effectiveness of entomopathogenic agents can be influenced by ambient temperature and humidity [12].

4 Conclusion

This study showed that despite the application of *L. muscarium* under severe climatic conditions in southern Tunisia, this fungus seems to be more effective against eggs and L2 larvae of *B. tabaci.* Thus, this fungus can be integrated as a means of microbiological control in an integrated management program (IPM) in heated greenhouses.

References

1. Landa Z, Osborne L, Lopez F, Eyal J (1994) A bioassay for determining pathogenicity of entomogenous fungi on whiteflies1. Biol Control 4(4):341–350. https://doi.org/10.1006/bcon.1994.1043
2. Gerling D, Mayer RT (1996) Bemisia: 1995 taxonomy, biology, damage, control and management. Intercept Limited
3. Alegbejo MD (2001) Whitefly transmitted plant viruses in Nigeria. J Sustain Agric 17(2–3):99–109. https://doi.org/10.1300/J064v17n02_10
4. Gilbertson RL, Batuman O, Webster CG, Adkins S (2015) Role of the insect super vectors *Bemisia tabaci* and *Frankliniella occidentalis* in the emergence and global spread of plant viruses. Annu Rev Virol 2(1):67–93
5. Jiang WJ, Peng YF, Ye JY, Wen YY, Liu GX, Xie JQ (2020) Effects of the entomopathogenic fungus *Metarhizium anisopliae* on the mortality and immune response of *Locusta migratoria.* Insects 11:36. https://doi.org/10.3390/insects11010036
6. Mascarin GM, Kobori NN, Quintela ED, Delalibera I Jr (2013) The virulence of entomopathogenic fungi against *Bemisia tabac*i biotype B (Hemiptera: Aleyrodidae) and their conidial production using solid substrate fermentation. Biol Control 66(3):209–218. https://doi.org/10.1016/j.biocontrol.2013.05.001
7. Sain SK, Monga D, Kumar R, Nagrale DT, Kranthi S, Kranthi KR (2019) Comparative effectiveness of bioassay methods in identifying the most virulent entomopathogenic fungal strains to control *Bemisia tabaci* (Gennadius) (Hemiptera: Aleyrodidae). Egypt J Biol Pest Control 29(1):1–11. https://doi.org/10.1186/s41938-019-0130-z
8. Cuthbertson AG, Walters KF, Northing P (2005) The susceptibility of immature stages of *Bemisia tabaci* to the entomopathogenic fungus *Lecanicillium muscarium* on tomato and verbena foliage. Mycopathologia 159(1):23–29. https://doi.org/10.1007/s11046-004-3127-3
9. Assadi B, Chouikhi S, Ettaib R, Laswad M, Aoun F, Belkhadi MS (2018) Etude de l'effet du champignon entomopathogène *Lecanicillummuscarium* contre *Bemisia tabaci* sous serres géothermiques dans le sud tunisien. Actes du 6ème Meeting International "Agriculture Oasienne et Développement Durable"
10. Assadi BH, Chouikhi S, Ettaib R, M'hamdi NB, Belkadhi MS (2021) Effect of the native strain of the predator *Nesidiocoris tenuis* Reuter and the entomopathogenic fungi *Beauveria bassiana* and *Lecanicillium muscarium* against *Bemisia tabaci* (Genn.) under greenhouse conditions in Tunisia. Egypt J Biol Pest Control 31(1):1–11. https://doi.org/10.1186/s41938-021-00395-5
11. Faria M, Wraight SP (2001) Biological control of *Bemisia tabaci* with fungi. Crop Prot 20(9):767–778. https://doi.org/10.1016/S0261-2194(01)00110-7
12. Mantzoukas S, Eliopoulos PA (2020) Endophytic entomopathogenic fungi: a valuable biological control tool against plant pests. Appl Sci 10(1):360. https://doi.org/10.3390/app10010360

Effect of Some Parameters on the Efficiency of Solar Distiller

Hiba Akrout, Khaoula Hidouri, Ali Benhmidene, and Bechir Chaouachi

Abstract This work aims to study the heat and mass transfer for the different constituents of a simple solar distiller. A comparison is made between simulation results and experimental measurements to validate the model. A parametric study has been then carried out to assess the influence of some parameters like the saline water sunlight absorption and depth, the contact angle of condensation surface and its thermal conductivity and the distiller height on the performance of solar still. The model results show that the production increases with the saline water absorption coefficient, while it decreases with basin water depth increase. In addition, the optimal contact angle for the condensation surface materials such as glass is between 0 and 40°. Results reveal that the glass production is two time greater than the case of polycarbonate. Moreover, it was concluded that the thermal conductivity of the condensation material has an insignificant effect on the distiller production. The solar still average height variation show that augmenting the distiller height for the same water depth and inclination angle, decreases its production.

Keywords Solar distillation · Contact angle · Thermal conductivity · Distiller · Height

1 Introduction

The world is facing a major water crisis due to drought and increasing consumption. The population is increasing rapidly and the water needs in the industrial and agricultural sectors are increasingly high. Water scarcity affects more than 40% of the population. Today there are many seawater desalination systems, the most popular ones are multiple effect distillation, multi stage flash distillation, vapor compression and use of membranes (reverse osmosis and electrodialysis). Desalination by solar

H. Akrout (✉) · K. Hidouri · A. Benhmidene · B. Chaouachi
Research Laboratory Energy, Water, Environment and Process, National Engineering School of Gabes, Gabes University, Gabes 6029, Tunisia
e-mail: akrouthiba@gmail.com

R. Khiari and M. Jawaid (eds.), *Proceedings of the 3rd International Congress of Applied Chemistry & Environment (ICACE-3)*, Springer Proceedings in Materials 23, https://doi.org/10.1007/978-981-99-1968-0_9

distillation is widely used in arid regions, on the scale of a small village or a family. The most widespread solar stills are those whose condensation surface is not separated from the evaporation and are called passive solar stills and stills whose condensation surface is separated from the heating of the water called active solar stills. The simple solar distillation produces water but with a low yield. Several theoretical and experimental studies have been made with the aim to increase the performance of distillation process. It was found that a Compressed air injected at different rates within the basin water of a hybrid solar distiller related to an air compressor enhanced the heat transfers rates and therefore the overall performance of the solar still [1]. Two rotating different discs with different rotation speeds were incorporated in the basin water of a simple solar with the aim to reduce the salty water thickness and improve the evaporation process. Results showed that the production of the rotating discs distillers are higher that without disc. Moreover, the optimum rotational speed was found to be 0.05 and 0.1 rpm, for both corrugated and flat discs with and without wick, respectively [2]. Lei Mu et al. increased the production of conventional solar still by about 467% by integrating a Fresnel lens to concentrate the sunlight. Results also showed that the system yield was enhanced when a cooling forced air is introduced [3]. A comparative study for a conventional solar still with different PCMS was elaborated by Kabeel et al. [4]. They reported that the capric–palmitic inorganic PCM and the organic PCM A48 are low-cost materials and ensures a high productivity. Moreover, they found that the thickness of PCM has an unsignificant effect on the solar still productivity. The effect of adding nanoparticles such as TiO_2, CuO, and GO to the basin phase change materials was investigated by Rufuss et al. [5]. The used phase change material was paraffin. Results revealed an increase in productivity by 26 and 35%, for solar stills with paraffin–TiO_2 and paraffin–CuO, respectively, compared to solar still with PCM only. The performance is influenced by parameters that are classified into internal and external [6]. In this paper, we present the results of a numerical parametric investigation of a simple solar still. The effect of parameters like the saline water sunlight absorption and depth, the cover contact angle and its thermal conductivity, the distiller height on the solar still production is evaluated.

2 System Description

The system studied is designed to benefit from the greenhouse effect. It is a closed distiller covered by a transparent cover. Under the greenhouse effect caused by the penetration of solar radiation and the nature of the cover, the brackish water placed in the basin will be heated. The heating process is initiated by the presence of a black galvanized iron absorber. The sides of the distiller are coated with a reflective material for a multiplication of the solar rays received by the liquid. To improve the performance of the studied system, a thermal insulation in glass wool is used.

The surface of the condenser exposed to the ambient air would condense the water vapor. A schematic of the studied distiller is shown in Fig. 1.

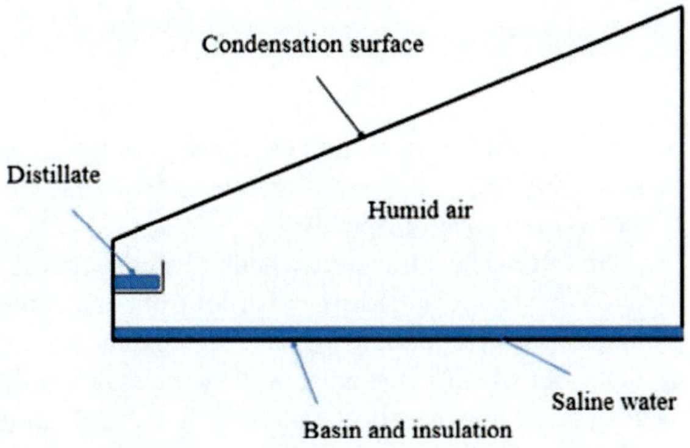

Fig. 1 Scheme of the studied system

3 Mathematical Model

The model was developed based on the energy balances of different components of the studied distiller such as the condensation surface, the water, the humid air, and the basin. A comparison between theoretical results and experimental measurements is then done to validate the model.

The internal surface of condensation heat balance

$$\frac{m_{ci}C_{pci}}{S_{ci}}\frac{dT_{ci}}{dt} = \alpha_{cc}G + \left(q_{e,ha-ci} + q_{c,ha-ci} + q_{r,ha-ci}\right) - q_{conduction,ci-cex} \quad (1)$$

The exterior surface of condensation heat balance

$$\frac{m_{cex}C_{pcex}}{S_{cex}}\frac{dT_{cex}}{dt} = q_{cond,ci-cex} - q_{r,cex-a} - q_{c,cex-a} \quad (2)$$

The saline water heat balance

$$\frac{m_{w}C_{pw}}{S_{w}}\frac{dT_{w}}{dt} = \tau_{cc}\tau_{dr}\alpha_{w}G + q_{c,b-w} - \left(q_{e,w-ha} + q_{c,w-ha} + q_{r,w-ha}\right) \quad (3)$$

The humid air heat balance

$$\frac{m_{ha}C_{pha}}{S_{ha}}\frac{dT_{ha}}{dt} = \left(q_{e,w-ha} + q_{c,w-ha} + q_{r,w-ha}\right) - \left(q_{e,ha-ci} + q_{c,ha-ci} + q_{r,ha-ci}\right) \quad (4)$$

The basin heat balance

$$\frac{m_{b}C_{pb}}{S_{b}}\frac{dT_{b}}{dt} = \tau_{cc}\tau_{dr}\tau_{w}\alpha_{b}G - q_{c,b-w} - q_{losses\,b} \quad (5)$$

The distillate flow rate is

$$\dot{m} = \frac{q_{\mathrm{e,ha-ci}}}{L_{\mathrm{v}}} \tag{6}$$

α_{w}, α_{b}, α_{cc} are the absorption coefficients of water, basin, and condensation surface, respectively; τ_{w}, τ_{dr}, τ_{cc} are the transmittance of the water, the droplets and condensation cover, respectively.

The various heat transfers whether by convection, radiation, conduction, or evaporation inside and outside the distiller from one component to another are presented in the following equations where T is the temperature, k is the thermal conductivity, ρ is the density, S is the area, m is the mass, and L is the thickness, the indices c, r, and e are mean convection, radiation, and evaporation, w, ha, b, cc, ci, cex, a, and ss are water, humid air, basin, condensation material, inner condensation side, external condensation side, and ambient air, and sides surfaces respectively [6–10].

$$q_{\mathrm{c,w-ha}} = h_{\mathrm{c,w-ha}}(T_{\mathrm{w}} - T_{\mathrm{ha}}) \tag{7}$$

$$q_{\mathrm{e,w-ha}} = h_{\mathrm{m}}(\rho_{\mathrm{w}} - \rho_{\mathrm{ha}}) \tag{8}$$

$$q_{\mathrm{c,b-w}} = h_{\mathrm{c,b-w}}(T_{\mathrm{b}} - T_{\mathrm{w}}) \tag{9}$$

$$q_{\mathrm{e,ha-ci}} = h_{m'}(\rho_{\mathrm{ha}} - \rho_{\mathrm{ci}}) \tag{10}$$

$$q_{\mathrm{r,ha-ci}} = \varepsilon_{\mathrm{eff}}\sigma\left(T_{\mathrm{ha}}^4 - T_{\mathrm{ci}}^4\right) \tag{11}$$

$$q_{\mathrm{conduction,ci-cex}} = \left(\frac{K_{\mathrm{cc}}}{L_{\mathrm{cc}}}\right)(T_{\mathrm{ci}} - T_{\mathrm{cex}}) \tag{12}$$

$$q_{\mathrm{r,cex-a}} = \varepsilon_{\mathrm{cex}}\sigma\left(T_{\mathrm{cex}}^4 - T_{\mathrm{sky}}^4\right) \tag{13}$$

$$q_{\mathrm{c,cex-a}} = h_{\mathrm{c,cex-a}}(T_{\mathrm{cex}} - T_{\mathrm{a}}) \tag{14}$$

$$q_{\mathrm{losses\,(b)}} = U_{\mathrm{b}}(T_{\mathrm{b}} - T_{\mathrm{a}}) \tag{15}$$

The sky temperature is related to the ambient temperature by Zurigat and Abu-Arabi [9]:

$$T_{\mathrm{sky}} = T_{\mathrm{a}} - 6 \tag{16}$$

$q_{\mathrm{r,cex-a}}$ and $q_{\mathrm{c,cex-a}}$ are the external heat losses by radiation and convection from the condensation cover to the outside [7, 10]. $q_{\mathrm{losses\,(b)}}$ are the heat losses from basin to the atmosphere [10].

The different heat transfers coefficients by convection, conduction radiation, and evaporation inside and outside the distiller are given below [7, 8, 11]

$$h_{c,w-ha} = \frac{Nu K_{ha}}{x} \tag{17}$$

$$h_m = \frac{h_{c,w-ha}}{\rho_{ha} C_{pha} Le^{(1-n)}} \text{ with } n = \frac{1}{4} \text{ if } 10^4 < Gr < 3 \times 10^5 \tag{18}$$

$$n = \frac{1}{3} \quad \text{if } 3.2 \times 10^5 < Gr < 10^7$$

$$h_{c,b-w} = \frac{0.54\left(K_w Ra_w^{\frac{1}{4}}\right)}{L_w} \quad \text{if } 10^4 < Ra_w < 10^7 \tag{19}$$

$$h_{c,b-w} = \frac{0.15\left(K_w Ra_w^{\frac{1}{3}}\right)}{L_w} \quad \text{if } 10^7 < Ra_w < 10^{11} \tag{20}$$

$$h_{c,ha-ci} = \frac{Nu K_{ha}}{x} \tag{21}$$

$$h_{m'} = \frac{h_{c,ha-ci}}{\rho_{ha} C_{pha} Le^{(1-n)}} \quad \text{with } n = \frac{1}{4} \quad \text{if } 10^4 < Gr < 3 \times 10^5 \tag{22}$$

$$n = \frac{1}{3} \quad \text{if } 3.2 \times 10^5 < Gr < 10^7$$

$$h_{c,cex-a} = 6.15 \times V_{wind}^{0.8} \quad \text{if } V_{wind} > 5 \text{ m/s} \tag{23}$$

$$h_{c,cex-a} = 2.8 + 3 \times V_{wind} \quad \text{if } V_{wind} \le 5 \text{ m/s} \tag{24}$$

$$U_b = \frac{1}{\left(\frac{L_b}{K_b} + \frac{L_i}{K_i} + \frac{1}{h_{c,b-a}}\right)} + \frac{S_{ss}}{S_b} \frac{1}{\left(\frac{L_b}{K_b} + \frac{L_i}{K_i} + \frac{1}{h_{c,b-a}}\right)} \tag{25}$$

h_m and $h_{m'}$ are the convective mass transfer coefficients from water to humid air and from humid air to the inner condensation cover, respectively [8].

The Nusselt number in Eqs. (17) and (21) is estimated by Mull and Reicher [12]:

$$Nu = 0.21(Gr\ Pr)^{1/4} \quad \text{if } 10^4 < Gr < 3 \times 10^5 \tag{26}$$

$$Nu = 0.075(Gr\ Pr)^{1/3} \quad \text{if } 3.2 \times 10^5 < Gr < 10^7 \tag{27}$$

These equations are appropriate for the evaluation of free convection. For the solar distiller, a modified Grashof number was proposed to estimate the simultaneous convection and mass transfer [13]. The Grashof numbers for the estimation of convection heat transfers from water to humid air and from humid air to the inner condensation side are given respectively by

$$\mathrm{Gr_m} = \frac{x^3 \rho_{\mathrm{ha}}^2 \beta_{\mathrm{ha}} g}{u_{\mathrm{ha}}^2} \left[(T_{\mathrm{w}} - T_{\mathrm{ha}}) + \frac{(P_{\mathrm{w}} - P_{\mathrm{ha}})(T_{\mathrm{w}} + 273.15)}{268.9 * 10^3 - P_{\mathrm{w}}} \right] \tag{28}$$

$$Gr'_{\mathrm{m}} = \frac{x^3 \rho_{\mathrm{ha}}^2 \beta_{\mathrm{ha}} g}{u_{\mathrm{ha}}^2} \left[(T_{\mathrm{ha}} - T_{\mathrm{ci}}) + \frac{(P_{\mathrm{ha}} - P_{\mathrm{ci}})(T_{\mathrm{ha}} + 273.15)}{268.9 * 10^3 - P_{\mathrm{ha}}} \right] \tag{29}$$

4 Model Results and Discussions

A parametric investigation is subsequently conducted where the effect of the variation of different parameters on the system performance is analyzed. The results of the parametric study will be presented in the following subsections.

4.1 *The Saline Water Sunlight Absorption*

Figure 2 illustrates the variation of production with saline water absorption coefficient. The increase of water absorption enhances its rate of evaporation and consequently the solar distiller production. On the experimental level, the water sunlight absorption coefficient may be enhanced, for example, by adding dye to the saline water [14].

4.2 *The Saline Water Depth*

As it's shown in Fig. 3 the total production is inversely proportional to the water thickness. For June month, increasing the water thickness from 0.5 to 3 cm, the daily distillate water is reduced by 630 mL/m^2 per day. In fact, the increase in water mass requires more heat to its evaporation.

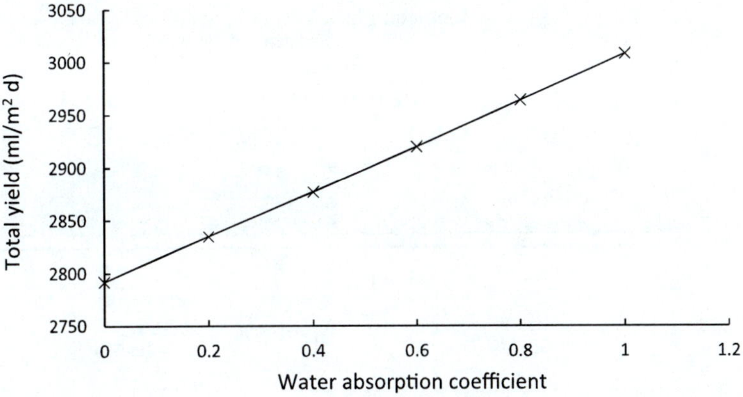

Fig. 2 Cumulative production versus water absorption coefficient

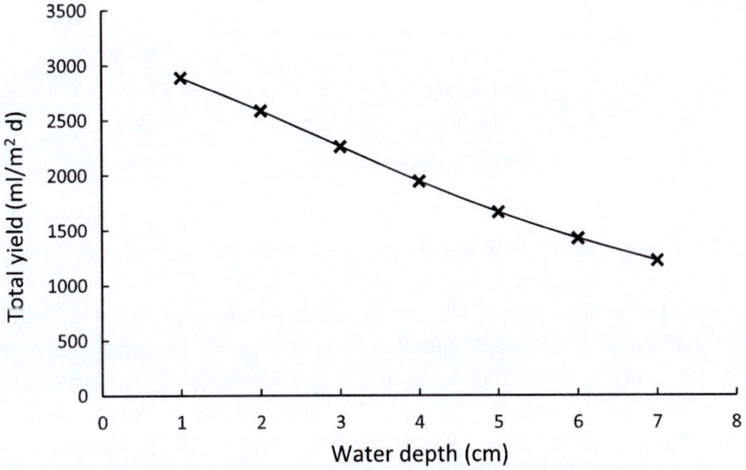

Fig. 3 Cumulative production with water depth

4.3 *The Choice of Condensation Surface Material*

Figure 4 illustrates the change in water droplet shape in function of contact angle that varies according to the chosen material. For contact angles less or equal to 40°, the droplet transmittance achieves 90%. An important decrease in transmittance happens, beyond 40°. It has the smallest value for 90° [15].

The sunlight transmittance has a significant effect on the still production; the material contact angle must not exceed 40° such as the glass. Thus, the glass material has the greatest cumulative production compared to other ones as presented in Table 1. According to the same table, the total production reaches 2438, 2810, 4614, and

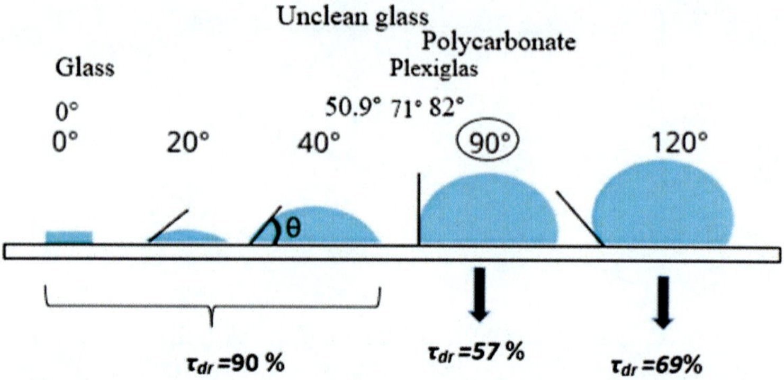

Fig. 4 Water droplet geometry versus contact angle

Table 1 Total productions for different condensation materials

	Contact angle (°)	Cumulative production (mL/m^2 day)
Glass	0	5058
Plexiglass	71	2810
Unclean glass	50.9	4614
Polycarbonate	90	2483

5058 mL/m^2 per day, respectively, for polycarbonate, plexiglass, unclean glass, and glass.

Moreover, the influence of the condensation surface thermal conductivity on distiller performance has been numerically investigated. Figure 5 shows that the cumulative distiller production is practically constant for different thermal conductivities for fixed contact angles of 0 and 90°. Therefore, the model results reveal that the thermal conductivity has no weighty effect on the global production compared to the contact angle effect discussed previously.

4.4 The Distiller Height

The numerical variation of the production for different solar still heights is shown in Table 2. Augmenting the height from 0.25 to 0.55 m, for the same basin surface, same inclination angle, and constant water depth leads to the production decrease by 247 and 383 mL/day for plexiglass and glass, respectively. An increase of the lateral surface area caused by the increase of the average height of the distiller has the effect to increase the external lateral heat losses.

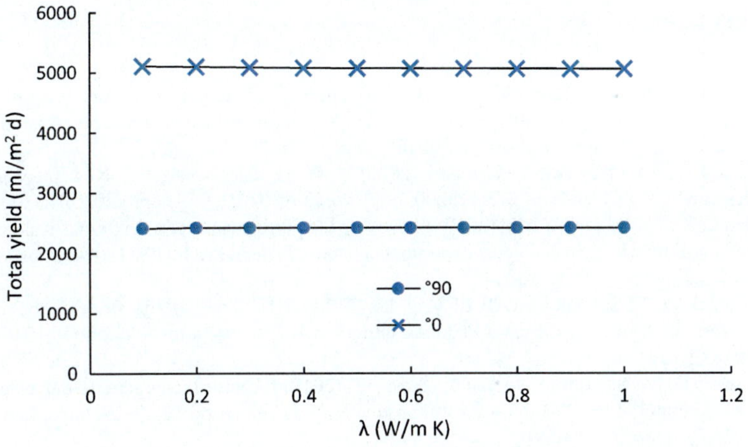

Fig. 5 Total production as a function of thermal conductivity for fixed contact angles

Table 2 Glass and Plexiglass cumulative productions for different distiller average heights

x	$V_{\text{plexiglass}}$ (mL/day)	V_{glass} (mL/day)
0.25	2895	5230
0.35	2810	5098
0.45	2727	4971
0.55	2648	4847
0.65	2571	4728

5 Conclusion

In this work, a mathematical model based on heat and mass balance equations for various solar still elements was carried out. A parametric study is then elaborated. Numerical simulations of typical day of June reveals that the production is getting higher when increasing the saline water absorption coefficient. However, the production is inversely related to the saline water depth. Moreover, the optimal contact angle for the condensation materials such as glass materiel is between 0 and 40°.

The glass production is two time greater than the case of polycarbonate. The total yield remains practically constant against the thermal conductivity for 0 and 90° contact angles. Therefore, the contact angle is the main condensation material property that affects the solar still production while the thermal conductivity change has no influence on the distiller performance.

References

1. Hidouri K, Mishra DR, Benhmidene A, Chouachi B (2017) Experimental and theoretical evaluation of a hybrid solar still integrated with an air compressor using ANN. Desalination Water Treat 88:52–59. https://doi.org/10.5004/dwt.2017.21333
2. Essa FA, Abdullah AS, Omara ZM (2020) Rotating discs solar still: New mechanism of desalination. J Clean Prod 275:123200. https://doi.org/10.1016/j.jclepro.2020.123200
3. Mu L, Xu X, Williams T et al (2019) Enhancing the performance of a single-basin single-slope solar still by using Fresnel lens: experimental study. J Clean Prod 239:118094. https://doi.org/10.1016/j.jclepro.2019.118094
4. Kabeel AE, El-Samadony YAF, El-Maghlany WM (2018) Comparative study on the solar still performance utilizing different PCM. Desalination 432:89–96. https://doi.org/10.1016/j.desal.2018.01.016
5. Rufuss DDW, Suganthi L, Iniyan S, Davies PA (2018) Effects of nanoparticle-enhanced phase change material (NPCM) on solar still productivity. J Clean Prod 192:9–29. https://doi.org/10.1016/j.jclepro.2018.04.201
6. Halloufi O (2010) Etude de la performance d'un distillateur solaire par un système de préchauffage solaire de l'eau saumatre
7. Abu-Arabi M, Zurigat Y, Al-Hinai H, Al-Hiddabi S (2002) Modeling and performance analysis of a solar desalination unit with double-glass cover cooling. Desalination 143:173–182. https://doi.org/10.1016/S0011-9164(02)00238-2
8. Hongfei Z, Xiaoyan Z, Jing Z, Yuyuan W (2002) A group of improved heat and mass transfer correlations in solar stills. Energy Convers Manag 43:2469–2478. https://doi.org/10.1016/S0196-8904(01)00185-6
9. Zurigat YH, Abu-Arabi MK (2004) Modelling and performance analysis of a regenerative solar desalination unit. Appl Therm Eng 24:1061–1072. https://doi.org/10.1016/j.applthermaleng.2003.11.010
10. Aybar HŞ, Assefi H (2009) Simulation of a solar still to investigate water depth and glass angle. Desalination Water Treat 7:35–40. https://doi.org/10.5004/dwt.2009.69
11. Madhlopa A, Johnstone C (2009) Numerical study of a passive solar still with separate condenser. Renew Energy 34:1668–1677. https://doi.org/10.1016/j.renene.2008.12.032
12. Mull W, Reicher M (1930) Gesundheit Beihoffe. Reihe 1:28
13. Sharpley BF, Boelter LMK (1938) Evaporation of water into quiet air from a one-foot diameter surface. Ind Eng Chem 30:1125–1131
14. Sodha MS, Kumar A, Tiwari GN, Pandey GC (1980) Effects of dye on the performance of a solar still. Appl Energy 7:147–162. https://doi.org/10.1016/0306-2619(80)90055-0
15. Briscoe BJ, Galvin KP (1991) The effect of surface fog on the transmittance of light. Sol Energy 46:191–197. https://doi.org/10.1016/0038-092X(91)90063-3

Reuse of Grain Waste as Abrasive Particles

Elaissi Arwa, Ghith Adel, and Alibi Hamza

Abstract We can observe that the use of abrasives has increased in recent years. In this study, the grain of perlites and pumice stones was harvested from industries and used as abrasive grains. From the tests carried out on these grains, it is possible to determine the application of the abrasives. The particles were sieved, cleaned, and sorted according to the particle size analysis which consists in determining the proportion of the different particle size classes. To determine the appearance of granules and to determine their diameter, scanning electron microscopy (SEM) was used. With both hollows and bumps present, the morphology of grains is very diverse. Pumice grains have more calcium atoms than perlite grains, according to investigations done using the inductively coupled plasma mass spectrometry (ICP/MS) technique. The perlite grains exhibit low resistance to wear in the micro-Deval test (MDE), whereas the pumice stone grains exhibit strong resistance.

Keywords Grains · Perlite · Pumice stones · Abrasives

1 Introduction

Surface flaws and extreme roughness are generally present in every industrially manufactured product, which affects the material's fatigue qualities. It, therefore, requires a significant shrinkage of material to obtain the desired surface state [1]. To remove some of the material's microscopic particles, this material removal procedure

E. Arwa (✉) · G. Adel
Textile Materials and Research Process (MPTEX), Tunisia National Engineering School of Monastir (ENIM), University of Monastir, Monastir, Tunisia
e-mail: earwa.elaissi@enim.u-monastir.tn

G. Adel
e-mail: ADEL.GHITH@u-monastir.tn

A. Hamza
Laboratory of Study of the Thermal and Energy Systems, National School of Engineering (ENIM), University of Monastir, Monastir, Tunisia

R. Khiari and M. Jawaid (eds.), *Proceedings of the 3rd International Congress of Applied Chemistry & Environment (ICACE-3)*, Springer Proceedings in Materials 23, https://doi.org/10.1007/978-981-99-1968-0_10

generates a lot of friction between the abrasive grains and the surface with which they come into contact. The application differs depending on the grain sizes. The roughing activities require coarse grains for a considerable removal of material. In contrast, the finishing work requires the use of fine grains. Kim et al. studied frictional behavior via a pin-on-disk wear test for aluminum alloys [2]. The size and type of abrasives have the greatest impact on surface morphologies and roughness [3].

Two characteristics are important to the final results of these products. The effectiveness of these products depends on two factors. The ultimate surface condition is determined by the hardness that allows for assault on softer materials and the size of the grains. In the investigation by Yang Chen et al., the PS/CeO_2 composite particles were made of amorphous PS polystyrene and cubic fluorite CeO_2 to increase chemical mechanical polishing (CMP) performance [4].

2 Material and Methods

2.1 *Materials Preparation*

Pumice stones and perlite were the two types of grains that were gathered for this investigation:

- Pumice stone is a low-density volcanic rock that is exceedingly porous. It is used to simulate fake wear and aging of stonewashed jeans due to their abrasiveness.
- Perlite is a volcanically formed natural rock. This thin, inert substance is non-combustible and ideal for a range of applications, including building and cosmetics. It is comfortable to wear.

According to the appropriate grain sizes, the grains are cleaned, sorted, crushed, and then filtered via sieves. In Monastir, Tunisia, SOTUCID SARL collaborated with researchers to complete the study.

2.2 *Characterization Using Scanning Electron Microscopy*

A method frequently used for examining the morphology and surface quality of materials is scanning electron microscopy (SEM). To evaluate the various features of the composites and examine the interactions between the reinforcement and the matrix before and after polishing, the specimens were scanned with an FEI Q250 Thermo Fisher-type scanner operating at 25 kV with various magnifications [5]. This result is provided by Thermo Fisher and analyzed in Tunisia National Engineering School of Monastir (ENIM, Tunisia). The sample is metalized to conduct this investigation, which makes the sample more conducive by coating it with a gold layer. A cathode beam of light is then used to bombard the sample with electrons once it has been evacuated. Signals are created as a result of the sample's interaction with the beam and will be captured and amplified to create the image.

2.3 ICP/MS Characterization

The determination of heavy metals is defined by inductively coupled plasma mass spectrometry. It is determined using the ISO 16711-1 standard in conjunction with the method for the determination of metals by microwave mineralization [6].

To identify the components of these materials, abrasive grains are subjected to this characterization. The in-line isotope dilution method was used for all measurements on a quadrupole ICP-MS (Agilent Technologies, 7700 ICP-MS series) at CETTEX (Technical Center of Textile in Tunisia). According to the manufacturer's instructions, the samples must first be prepared for each recipient before being placed in the microwave. For metallic products, this entails adding 9 mL of concentrated nitric acid (HNO_3) and 3 mL of concentrated hydrochloric acid to each 100 mL container before submerging the mass sample (HCl). For non-metallic items, samples are crushed to speed up the dissolution process before being mixed with 6 mL of concentrated nitric acid (HNO_3) and 2 mL of concentrated hydrochloric acid in each 100 mL container (HCl). Additionally, certified reference content has been added. After setting up the microwave digesting system, the mixture was put into the microwave oven. The contained containers are then transferred into a 50 mL volumetric flask. After that, calibration solutions made from the proper matrix are used to apply the metal assay up to the gauge line. The outcomes are the mass concentration of every metal present in the substance.

The following formula is used to determine the metal content [7]:

$$d = \frac{\beta \times V}{m \times 1000} \tag{1}$$

with

d: the metal content (mg/kg)
β: the mass concentration of metals (µg/L)
V: the volume of solvent (mL)
m: mass of the test piece (g).

2.4 Micro-Deval Wear Characterization

The micro-Deval test, which is based on European Standard EN 1097-1, is used to evaluate an aggregate's resistance to wear caused by reciprocal friction. The friction between the components, on the cylinder of the revolving machine, and the balls (abrasive load), whether dry or wet, causes the material to alter throughout the test.

A sample with a mass of at least 2 kg is dried at 105 °C in the Ibn Sina Sectorial Building Training Center (CSFBA, Tunisia) until two subsequent weighings that are separated by an hour do not differ by 0.10%. The produced sample is then loaded with 2500 mL of tap water and 2000 g of steel balls in a steel drum [8]. The materials are then rotated 12,000 times at a speed of 100 revolutions per minute in a drum with an interior diameter of 200 mm. Following polishing, the material is rinsed and put through a 1.6 mm filter. An 8 mm discharge sieve will keep the abrasive charge. The residue is dried to a consistent mass at 1.6 mm in an oven at 105 °C before being weighed [9]. The following formula [10] is used to determine the Micro-Deval coefficient (MDE) [10]:

$$\mathrm{MDE} = \frac{m}{M} \times 100 \tag{2}$$

where $m = 500 - m_1$ and $M = 500$ g.

3 Results and Discussions

3.1 *Scanning Electron Microscopy of Grains*

According to Fig. 1a, b, the perlite grains are around 12 μ in size and the pumice stone grains range from 280 to 290 microns. With both hollows and bumps present, the morphology of grains is very diverse. Because there are so many microscopic grains, abrasive grains have a lot of sharp edges [11].

Pumice stone and perlite grains are divided into two sizes as a consequence of the particle size examination. The tiny grains have a diameter of less than 0.65 mm, as long as the big granules are between 1 and 2 mm in diameter.

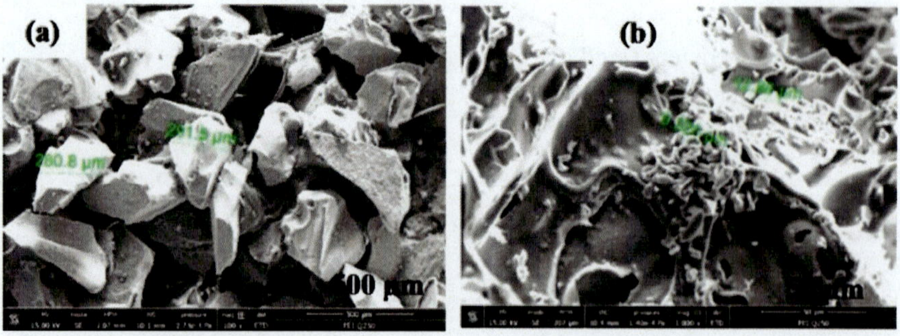

Fig. 1 SEM spectroscopy of **a** pumice stones and **b** perlite (500 μm)

Table 1 ICP-MS results of grains

Chemical element	Pumice stones quantity (g/kg)	Perlites quantity (g/kg)
^{27}Al	19.383	16.992
^{39}K	80.766	71.524
^{44}Ca	296.286	192.638
^{51}V	0.240	0
^{52}Cr	6.265	3.559
^{56}Fe	38.460	40.459
^{60}Ni	0.738	0.689
^{111}Cd	11.403	11.384

3.2 ICP/MS

Table 1 shows that the grains of pumice stones and perlites mainly contain calcium about 65% and 57% respectively and potassium about 18 and 21% respectively. Small amounts of the following metals are also found: Al, Cr, Fe, Ni, and Cd. For pumices stones, there is also a small amount of vanadium.

3.3 Wear Resistance Result (Micro-Deval)

This analysis is carried out for the coarse grains of pumice stones and perlites. Based on Table 2, the perlite grains exhibit poor resistance (MDE = 31%), whereas the pumice stone grains exhibit good resistance (MDE = 4 < 10%) [12].

Table 2 Type of grains according to the MDE coefficient

Micro-Deval coefficient values in the presence of water (%)	Assessment
<10	Very good to good
From 10 to 20	Good to medium
20–35	Medium to low
>35	Mediocre

4 Conclusion

Pumice stone and perlite particles have been used as abrasive grains because of their characteristics. The ICP/MS analysis specifies the co-components of the grains. SEM images show a few irregularities in composite surfaces.

We can categorize the created abrasives according to the item to be polished and the desired destination depending on the future demands of composite applications (finishing, polishing, etc.).

References

1. Rowe WB (2013) Principles of modern grinding technology. William Andrew
2. Kim H-J, et al (2005) Effects of tribooxidation on sliding behavior of aluminum. Wear 259(1–6):501–505
3. Guo J, et al (2017) Experimental investigation into the effect of abrasive and force conditions in magnetic field-assisted finishing. Int J Adv Manuf Technol 90(5–8):1881–1888
4. Chen Y, Lu J, Chen Z (2011) Preparation, characterization, and oxide CMP performance of composite polystyrene-core ceria-shell abrasives. Microelectron Eng 88(2):200–205
5. Jebril S, Jenana RKB, Dridi C (2020) Green synthesis of silver nanoparticles using Melia azedarach leaf extract and their antifungal activities: in vitro and in vivo. Chem Phys Mater 248:122898
6. Thoden van Velzen EU, et al (2020) Effect of recycled content and rPET quality on the properties of PET bottles. Part II: migration. Packag Technol Sci 33(9):359–371
7. Lewinska A (2018) Results of proficiency test metals in dried paint
8. Wang D, et al (2015) Evaluation of aggregate resistance to wear with micro-Deval test in combination with aggregate imaging techniques. Wear 338(288–296)
9. Augustin H, Eppensteiner W, Zieger M (1999) Verschleißversuche an Straßenbaugesteinen nach EN 1097-1. Strassenforschung 494
10. Gökalp İ, Uz VE, Saltan M (2016) Testing the abrasion resistance of aggregates including by-products by using micro Deval apparatus with different standard test methods. Constr Build Mater 123(1–7)
11. Huang B, et al (2021) Advances in fabrication of ceramic corundum abrasives based on sol–gel process. Chin J Aeronaut 34(6):1–17
12. Standard N (1996) Tests for mechanical and physical properties of aggregates. Part

Fast Removal of Methyl Parathion from Aqueous Solution Using Rice Husk Ash

Abdoulaye Demba N'Diaye, Noureddine Baaka, Mohamed Kankou, and Hatem Dhaouadi

Abstract This work investigated the possibility of using Rice Husk Ash (RHA) to remove the Methyl parathion (MP) pesticide from aqueous solution. The adsorbent dosage and the contact time impacts were studied by adsorption in batch procedures. The kinetic models were applied. The Langmuir, Freundlich and Redlich–Peterson isotherms were applied to the adsorption equilibrium data. Results demonstrate that the parameters of the isotherm models are adequate to fit the presented adsorption isotherm data in the order of: Redlich–Peterson > Freundlich > Langmuir. The Redlich–Peterson isotherm exponent n is less than unity indicated that the adsorption data were more of Freundlich. The maximum adsorption capacity, q_{m}, was found to be 8.95 mg g^{-1}.

Keywords Pesticide · Methyl parathion · Rice husk ash · Adsorbent

1 Introduction

The quasi-chronic contamination of natural waters by pesticides is not a novelty, but remains a current problem, whether for the public authorities, for water suppliers or for consumers.

The term pesticide derives from the English word "pest" which designates any animal, plant or microorganism (virus, bacteria, fungus, worm, mollusc, insect, bird, rodent and mammal) likely to be harmful to humans or to the environment [7].

In the environmental field, a pesticide is any commercial product or any chemical or biological substance other than a drug or a vaccine, used against a pest (insect,

A. D. N'Diaye (✉) · N. Baaka · H. Dhaouadi
Faculty of Sciences of Monastir, Research Laboratory—Environmental Chemistry and Clean Processes (LR21ES04), University of Monastir, 5000 Monastir, Tunisia
e-mail: abdouldemba@yahoo.fr

A. D. N'Diaye · M. Kankou
UR Eaux Pollution et Environnement, Département de Chimie, Faculté des Sciences et Techniques, Université de Nouakchott Al Aasriya, Nouakchott, Mauritanie

R. Khiari and M. Jawaid (eds.), *Proceedings of the 3rd International Congress of Applied Chemistry & Environment (ICACE-3)*, Springer Proceedings in Materials 23, https://doi.org/10.1007/978-981-99-1968-0_11

fungus, bacterium, virus, weed, rodent, ...), i.e. any living organism harmful, noxious or disturbing to humans and their resources, with the aim of eliminating it or inhibiting its development.

Organophosphates are universally used as pesticides in agriculture [6]. Of the organic phosphate compounds, methyl parathion (MP) is commonly used in wheat, rice and sugar crops. Several toxicological studies have been conducted on MP [27]. The World Health Organization has classified MP as a "highly hazardous" Class IA pesticide [12].

A number of treatment methods are utilized to eliminate MP from aqueous solution such as microbiological treatment [21], photocatalytic treatment [31], photolytic treatment [28], electrochemical oxidation [2] and adsorption treatment [20].

The adsorption is a surface phenomenon by which atoms or molecules of a fluid (adsorbates) are fixed on a solid surface (adsorbent) according to various processes more or less intense thanks to physical and/or chemical interactions [26]. Adsorption is one of the techniques very frequently adopted nowadays for the treatment of water and industrial effluents.

Its principle is based on the property that solids have to fix on their surfaces some pollutants on their surfaces [18].

The term adsorption was first proposed by Kayser in 1881 to differentiate between the condensation of gas on the surface, and a gas absorption, a process in which gas molecules penetrate the mass [23]. The phenomena of which adsorption were initially observed in liquids by Lo witz in 1785, and the first industrial application was realized a few years later in a sugar cane refinery to decolorize syrups. In 1860, charcoal was used to remove taste and odour from water. Subsequently, at the beginning of the twentieth century, adsorption techniques were developed thanks to scientific knowledge.

According to the types and nature of the adsorbate-adsorbent interactions or forces that adsorbates on the solid surface, we distinguish two types of adsorptions: physisorption (physical adsorption or physisorption (physical adsorption or Van Der Waals adsorption) and chemisorption (chemical adsorption or activated adsorption) [22].

The First World War saw the appearance of silica gels, then, in the years 1939–1940, activated aluminas. In 1945, the exceptional adsorption properties of natural zeolites were of natural zeolites. In 1950, the first synthetic zeolites paved the way for the fantastic development of molecular sieves as catalysts and adsorbents. In addition to these adsorbents, which are used in industrial quantities, new products with better properties have been developed in recent years. Only adsorbents with sufficient specific surface area can be of practical interest. Industrial adsorbents generally have specific surface areas of more than 100 m^2/g and up to several thousand m^2/g [3]. These adsorbents are necessarily microporous with pore sizes below 2 nm or mesoporous with pore sizes between 2 and 50 nm. Adsorbents are characterized by their external properties such as their specific surface area or their polarity. A large specific surface area is preferable for a high adsorption capacity. The size of the micropores determines the accessibility of the adsorbable molecules to the internal adsorption surface, so it is possible to characterize adsorbents by the pore size distribution, and thus to choose a particular adsorbent for a particular separation [29].

The adsorption technique is the most effective for the treatment of pesticides in the aquatic medium [10]. In addition, the commercial activated carbon is mainly used as adsorbent. Nevertheless, the cost of this product is extremely expensive.

In this context natural highly cost-effective nature and eco-friendly adsorbents are preferred. As a result, several inexpensive adsorbents on the removal of MP have been reported in the literature [1, 9, 14–17, 24].

The promising results of rice husk ash (RHA) when used as an adsorbent to eliminate atrazine [19] prompted the search for new methods to remove PM. It is the aim of this study, the adsorbent dosage and the effect of contact time have been investigated. To this end, kinetic and isotherm studies were performed using nonlinear methods.

2 Experimental

2.1 Adsorbent and Adsorbate

Rice husk collected from Mauritania was washed with ultrapure water, dried at 100 °C until reaching a constant weight and then ground. The RHA was obtained from burning rice husk in an electric oven at 400 °C for 3 h [13]. The RHA was sieved (<100 μm) and stored in a desiccator for further processing.

The moisture, ash and loss mass ignition values of the rice husk are 5.8%, 19.9% and 21.1%, respectively. The cellulose, lignin and hemicellulose values of the rice husk are 29.1%, 19.7% and 20.1%, respectively [19]. The results of the physical and chemical analysis of RHA showed pH_{pzc} of 6.8, total surface acidity of 0.177 m_{eq} g^{-1} and total surface basicity of 0.148 m_{eq} g^{-1}. The BET surface for RHA was found to be 147.6 m^2 g^{-1}.

Ultrapure water was used to wash the rice husks, collected from Mauritania. Then, dried at 100 °C to a constant weight and pulverized. The RHA was produced by cooking the rice husks in an electric oven at 400 °C for 3 h [13]. The RHA was filtered (<100 μm) and stockpiled in a desiccator for later treatment. Physical and chemical analysis results of RHA indicated a pH_{pzc} value of 6.8 and a total surface acidity of 0.177 m_{eq} g^{-1} and a total surface basicity of 0.148 m_{eq} g^{-1}. The BET surface area of RHA was found to be 147.6 m^2 g^{-1}. Figure 1 illustrates the MP molecule.

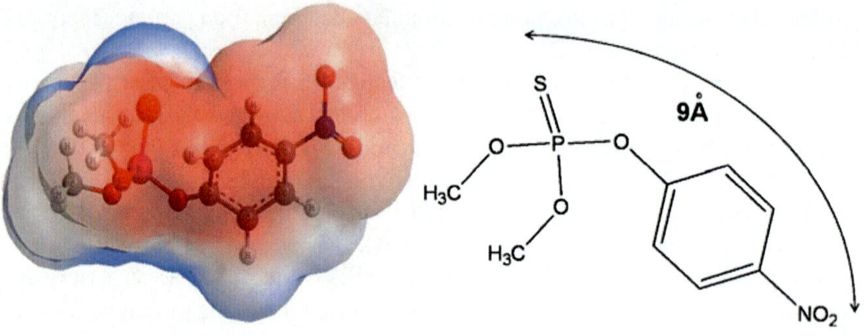

Fig. 1 MP chemical structure

2.2 Batch Experiments

PM Adsorption using the RHA as adsorbent was conducted in batch studies. In all series of experiments, fixed concentration of MP (5 mg L^{-1}) was stirred (70 rpm) with various doses of adsorbent for several periods of time. To explore the impact of contact time on the removal of PM by adsorption, 0.3 g of RHA was mixed with 25 mL of PM at 5 mg L^{-1} and stirred at 70 rpm at room temperature. Adsorption isotherms were obtained by varying the initial MP concentration from 5 to 100 mg L^{-1}. At the end of each experiment, the stirred solution mixture was micofiltered using micro filter and the residual concentration of MP was determined by HPLC.

The adsorption capacity (q_e) and percentage of MP removed R (%) was obtained using, respectively, the following Eqs. (1) and (2):

$$q_e = \frac{(C_i - C_e)V}{m} \quad (1)$$

$$R\ (\%) = \frac{C_i - C_e}{C_i} \times 100 \quad (2)$$

With

q_e (mg g^{-1}): amount of MP adsorbed by RHA,

C_i (mg L^{-1}): initial liquid-phase concentrations of MP,

C_e (mg L^{-1}): liquid-phase concentration of MP,

V (L): solution volume,

m (g): mass of RHA used.

2.3 Kinetic Study

In the purpose to perform the rate and the mechanism of the adsorption process, the pseudo-first order (PFO) model, pseudo-second order (PSO) model and Elovich model [8] were used. The non-linear PFO, PSO and Elovich model kinetics can be described by Eqs. (3), (4) and (5), respectively:

$$q_t = q_e\left(1 - \exp^{-k_1 t}\right) \tag{3}$$

$$q_t = \frac{k_2 q_e^2 t}{1 + k_2 q_e t} \tag{4}$$

$$q_t = \left(\frac{1}{\beta}\right) Ln(1 + \alpha\beta t) \tag{5}$$

With

q_t (mg g^{-1}): the amount of MP adsorbed per unit mass of RHA at time t,

k_1 (L mn^{-1}): the PFO rate constant,

k_2 (g mg^{-1} min^{-1}): the rate constant for adsorption,

q_e (mg g^{-1}),

t (min): the contact time,

α (mg g^{-1} min^{-1}): the initial adsorption rate,

β (g mg^{-1}): the extent of surface coverage and activation energy for chemisorption.

2.4 Equilibrium Adsorption Modelling

In this work, the Langmuir, Freundlich and Redlich–Peterson have been applied to study the equilibrium adsorption. The Langmuir adsorption isotherm assumes that the adsorption occurs on homogeneous specific surface sites in the adsorbent [30]. This is the most used model to comment on the results found during the adsorption of organic compounds in aqueous solution. The nonlinear Langmuir model can be expressed by Eq. (6):

$$q_e = \frac{q_m K_L C_e}{1 + K_L C_e} \tag{6}$$

With

q_e (mg g^{-1}): the amount of MP adsorbed per unit mass of RHA adsorbents,

K_L (M g^{-1}): the Langmuir constant related to the adsorption capacity,

C_e (mg L^{-1}): the concentration of MP in the solution at equilibrium,

q_m (mg g^{-1}): maximum uptake per unit mass of RHA adsorbents.

The nonlinear representation of the Freundlich model (Eq. (7)) is a widely used empirical equation for the practical representation of adsorption equilibrium. It has no theoretical basis. However, experience shows that it describes well the results of adsorption of micropollutants by solids such as activated carbons activated carbons, soils and clays [30]:

$$q_e = K_F C_e^{1/n} \tag{7}$$

With

K_F (mg g^{-1}) (L $mg^{-1})^n$: the constant related to adsorption capacity,

$1/n$: the constant related to adsorption intensity.

The non-linear representation of the Redlich–Peterson model is presented in Eq. (8) [5]:

$$q_e = \frac{K_{RP} C_e}{1 + \alpha_{RP} C_e^n} \tag{8}$$

With

K_{RP} (L g^{-1}): Redlich–Peterson isotherm constants,

α_{RP} (L mol^{-1}): Redlich–Peterson isotherm constants,

n: exponent, which lies between 0 and 1.

3 Results and Discussion

3.1 *Influence of Adsorbent Dosage*

The results of the removal of MP by RHA by varying mass in the range 0.1–2 g at fixed MP concentration 5 mg L^{-1} was presented in Fig. 2. The results given in Fig. 2 indicate that as the mass of the adsorbent increases, the adsorption increases due to the increase in surface area. However further increase over a certain mass will not increase the adsorption. Finally, for the further development of the kinetic models and adsorption isotherms, a mass of 0.3 g of RHA was used.

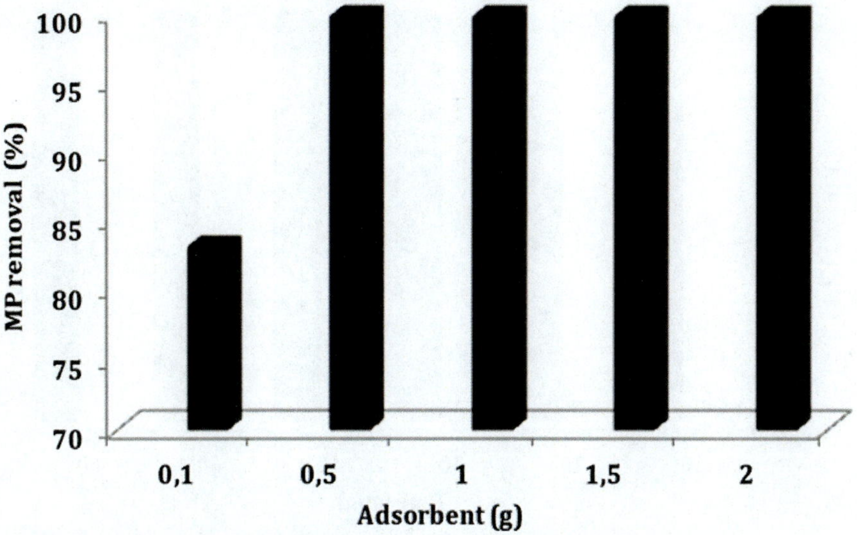

Fig. 2 Effect of adsorbents dosage on percentage adsorption of MP

3.2 *Kinetic Study*

The influence of the contact time (10–60 min) on removal adsorption of MP for RHA are presented in Fig. 3. We can observe that the equilibrium can be obtained at 30 min with a maximum removal efficiency equal to 98.4%. The short equilibrium time were reported by N'diaye et al. [20].

PFO, PSO and Elovich models are used to analyse the adsorption data kinetics. Results are presented in Fig. 4. Table 1 shows the values of various model parameters such as k_1, k_2, α, β and R^2.

Table 1 illustrates that the three studied kinetic models are appropriate in the following order: PSO > Elovich > PFO. From the R^2 values, it was also confirmed that the PSO model fits the adsorption data best, indicative of the chemisorption mechanism. The same phenomena have been previously reported for the adsorption of pesticides on several different adsorbents [19, 25].

On the other hand, Pearson's correlation matrix was used to explore the relationships between the kinetic models as shown in Table 2. The results demonstrated that the strong and positive correlation obtained between the PSO and the Elovich (r = 0.990) is superior to the positive correlation obtained between PFO and Elovich models (r = 0.924). The results confirmed that the equilibrium data were most appropriately represented by the PSO and Elovich models. According to Deokar and Mandavgane [4], the improved fit of the PSO and Elovich kinetic models to the experimental results suggests the chemisorption of the pesticide on the adsorbent surface.

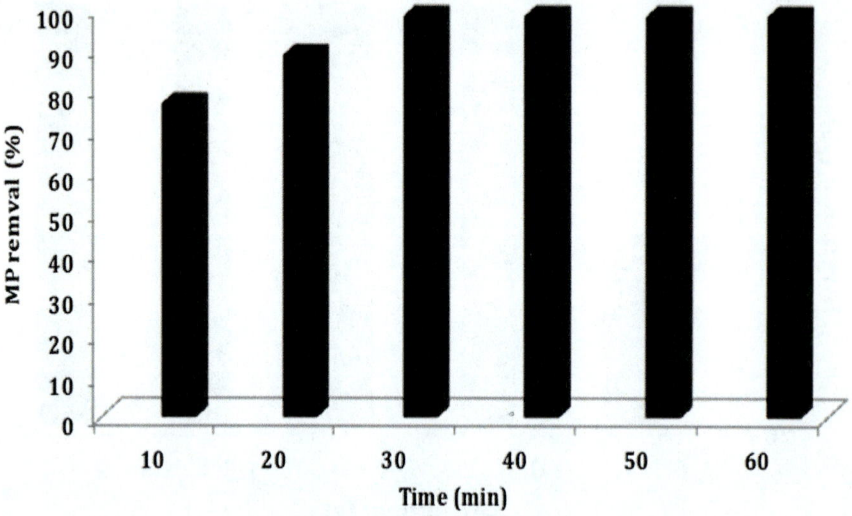

Fig. 3 Removal percentage of MP by RHA adsorbent

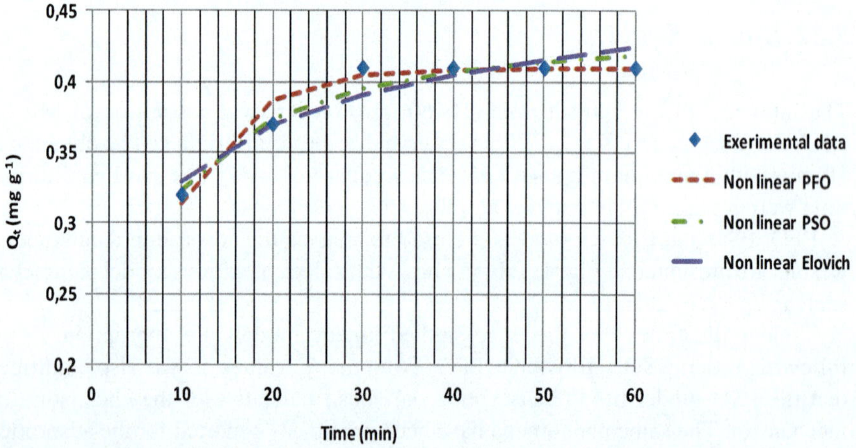

Fig. 4 PFO, PSO and Elovich non linear for RHA adsorbent

3.3 Adsorption Isotherms

Several laws have been proposed for the study of adsorption. They express the relation between the quantity adsorbed and the concentration of solute in a solvent at a given temperature.

The adsorption isotherm is the curve linking, at a fixed temperature, the quantity of Product adsorbed by initial mass of adsorbent (Q) to the concentration remaining in the fluid phase after obtaining the adsorption equilibrium (C_e).

Table 1 The constants for the adsorption of MP by RHA using the three models (PFO, PSO and Elovich)

Model	Parameters	Value
	q_{exp}	0.410
PFO	q_e	0.409
	k_1	0.145
	R^2 (%)	94.7
PSO	q_e	0.445
	k_2	0.59
	R^2 (%)	99.9
Elovich	α	0.57
	β	2.64
	R^2 (%)	97

Table 2 Correlation matrix of the kinetic models

Variables	PFO	PSO	Elovich
PFO	**1**		
PSO	0.969	**1**	
Elovich	0.924	0.990	**1**

The interest of the adsorption isotherm for an adsorbent/adsorbate system is multiple. It also allows to put forward hypotheses on the adsorption mode, indeed, its shape is representative of some phenomena involved: monolayer or multilayer adsorption. Lateral interactions between molecules or not.

The adsorption isotherm is used to investigate the maximum capacity of adsorption of PM on RHA. The obtained curves and the theoretical parameters of the isotherm are then compared with the measured data of the RHA adsorbent for PM removal in Table 3 and Fig. 5.

Table 3 Isotherm parameters for MP onto RHA

Models	Parameters	Value
Langmuir	q_m	8.94
	K_L	0.051
	R^2 (%)	98.36
Freundlich	$1/n$	0.60
	K_F	0.70
	R^2 (%)	99.38
Redlich–Peterson	K_{RP}	2.73
	α_{RP}	3.07
	n	0.45
	R^2 (%)	99.41

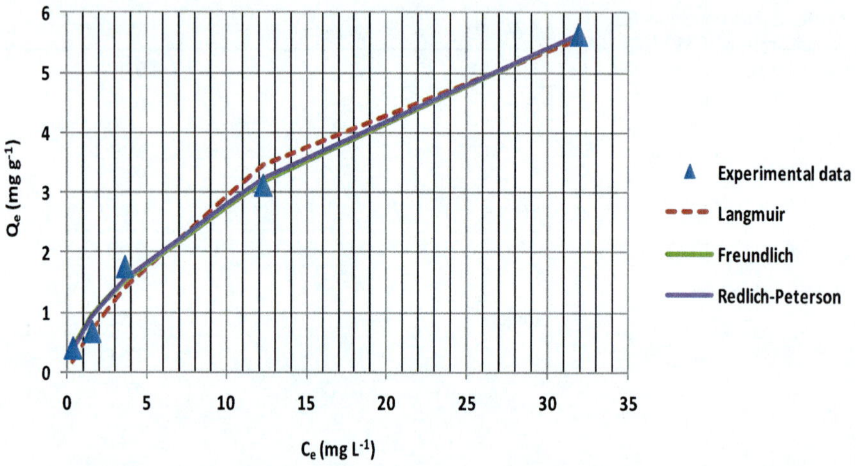

Fig. 5 MP adsorption by RHA: experimental and predicted isotherms

The results compiled in Table 3 indicate that the isotherm model parameters are fitted to the current adsorption isotherm data in the sequence Redlich–Peterson > Freundlich > Langmuir. The Redlich–Peterson isotherm exponent n is less than unity indicated that the adsorption data were more of Freundlich form suggesting that the surface of RHA is heterogeneous for the adsorption of the MP.

On the other hand, Pearson's correlation matrix was used to explore the relationships between the isotherm models (Table 4). Results presented in Table 4 confirmed that the Redlich–Peterson and Freundlich isotherms may well represent the experimental adsorption data ($r = 1$).

Adsorption capacity (q_m) values of selected PM adsorbents in the literature are shown in Table 5 for further comparison. The results of the comparison showed that (q_m) of RHA is found superior with many reported adsorbent.

Table 4 Correlation matrix of the isotherm models

Variables	Experimental data	Langmuir	Freundlich	Redlich–Peterson
Experimental data	**1**			
Langmuir	0.994	**1**		
Freundlich	0.997	0.996	**1**	
Redlich–Peterson	0.997	0.997	1.000	**1**

Table 5 q_m of RHA and selected adsorbents for MP adsorption

Adsorbents	q_m (mg g^{-1})	References
Sandy soil	0.84	[11]
Red soil	2.28	[11]
Clayey soil	3.49	[11]
Compost soil	6.45	[11]
Vermicompost	0.17	[17]
Typha australis	7.67	[20]
RHA	8.95	Present study

4 Conclusions

The MP adsorption kinetics by RHA was evaluated using a non-linear fitting curve approach to fit the PFO, PSO and Elovich equations. The obtained results showed that the PSO model fit the experimental data well. A non-linear method was used to study the equilibrium data, fitting them to the Freundlich, Langmuir and Redlich–Peterson model equations. In addition, the Redlich–Peterson isotherm exhibited an exponent n less than unity. This result indicates that the adsorption data are rather Freundlich in form suggesting that the RHA surface is heterogeneous for PM adsorption.

References

1. Akhtar M, Syed MH, Bhanger MI, Shahid I (2007) Low-cost sorbents for removal of methyl parathion pesticide from aqueous solution. Chemosphere 66(10):1829–1830
2. Alves SA, Ferreira TCR, Migliorini FL, Baldan MR, Ferreira NG, Lanza MRV (2013) Electrochemical degradation of the insecticide methyl parathion using a boron-doped diamond film anode. J Electroanal Chem 702:1–7
3. Darpentigny C, Nonglaton G, Bras J, Jean B (2020) Highly absorbent cellulose nanofibrils aerogels prepared by supercritical drying. Carbohydr polym 229:115560
4. Deokar SK, Mandavgane SA (2015) Rice husk ash for fast removal of 2,4-dichlorophenoxyacetic acid from aqueous solution. Adsorp Sci Technol 33:5
5. Dhaouadi H, M'Henni F (2009) Vat dye Sorption onto crude dehydrated sewage sludge. J Hazard Mater 164(2–3):448–458
6. Du D, Chen W, Zhang W, Liu D, Li H, Lin Y (2010) Covalent coupling of organophosphorus hydrolase loaded quantum dots to carbon nanotube/Au nanocomposite for enhanced detection of methyl parathion. Biosens Bioelectron 25(6):1370–1375
7. Feigenbrugel V, Le Calvé S, Mirabel P (2004) Temperature dependence of Henry's law constants of metolachlor and diazinon. Chemosphere 57:319–327
8. Guo L, Li G, Liu J, Meng Y, Xing G (2012) Nonlinear analysis of the kinetics and equilibrium for adsorptive removal of Cd (II) by starch phosphate. J Disper Sci Technol 33(3):403–409
9. Gupta VK, Gupta B, Rastogi A, Agarwal S, Nayak A (2011) Pesticides removal from waste water by activated carbon prepared from waste rubber tire. Water Res 45:4047–4055
10. Hameed BH, Salman JM, Ahmad AL (2009) Adsorption isotherm and kinetic modeling of 2,4-D pesticide on activated carbon derived from date stones. J Hazard Mater 163:121–126
11. Krishna RK, Philip L (2008) Adsorption and desorption characteristics of lindane, carbofuran and methyl parathion on various Indian soils. J Hazard Mater 160:559–567

12. Kumawat G, Gaur N, Karnawat R, Sharma IK, Verma PS (2016) Adsorption studies of methyl parathion on papaya seed activated carbon: an ecofriendly approach. World J Pharm Res 5(4):907–918
13. Mahvi AH, Maleki A, Eslami A (2004) Potential of rice husk and rice husk ash for phenol removal in aqueous systems. Am J Appl Sci 1(4):321–326
14. Memon GZ, Bhanger MI, Akhtar M (2007) The removal efficiency of chestnut shell for selected pesticides from aqueous solutions. J Colloid Interface Sci 315:33–40
15. Memon GZ, Bhanger MI, Akhtar M, Talpur FN, Memon JR (2008) Adsorption of methyl parathion pesticide from water using watermelon peels as a low-cost adsorbent. Chem Eng J 138(1–3):616–621
16. Memon GZ, Bhanger MI, Memon JR, Akhtar M (2009) Adsorption of methyl parathion from aqueous solutions using mango kernels: equilibrium, kinetic and thermodynamic studies. Bioremediat J 13(2):102–106
17. Mendes CB, Lima GF, Alves VN, Tarley CRT (2012) Evaluation of vermicompost as a raw natural adsorbent for adsorption of pesticide methyl parathion. Environ Technol 33(1–3):167–172
18. Mondol MMH, Jhung SH (2021) Adsorptive removal of pesticides from water with metal–organic framework-based materials. Chem Eng J 421:129688
19. N'diaye AD, Boudokhane C, Kankou M, Dhaouadi H (2019) Potential of rice husk ash in atrazine removal. Chem Ecol 1–15
20. N'diaye AD, Boudokhane C, Elkory MB, Kankou M, Dhaouadi H (2018) Methyl parathion pesticide removal from aqueous solution using Senegal River *Typha Australis*. Water Science and Technology. Water Supply 18(5):1545–1553
21. Pino N, Peñuela G (2011) Simultaneous degradation of the pesticides methyl parathion and chlorpyrifos by an isolated bacterial consortium from a contaminated site. Int Biodeterior Biodegrad 65(6):827–831
22. Puccia V, Avena MJ (2021) On the use of the Dubinin–Radushkevich equation to distinguish between physical and chemical adsorption at the solid-water interface. Colloid Interface Sci Commun 41:100376
23. Robens E, Jayaweera SAA (2014) Early history of adsorption measurements. Adsorp Sci Technol 32(6):425–442
24. Saha A, Gajbhiye VT, Gupta S, Kumar R, Ghosh RK (2014) Simultaneous removal of pesticides from water by rice husk ash: batch and column studies. Water Environ Res 86:11
25. Salman JM, Njoku VO, Hameed BH (2011) Bentazon and Carbofuran adsorption onto date seed activated carbon: kinetics and equilibrium. Chem Eng J 173:361–368
26. Sarkar B, Mukhopadhyay R, Mandal A, Mandal S, Vithanage M, Biswas JK (2020) Sorption and desorption of agro-pesticides in soils. In: Agrochemicals detection, treatment and remediation. Butterworth-Heinemann, pp 189–205
27. Suthar S (2013) Toxicity of methyl parathion on growth and reproduction of three ecologically different tropical earthworms. Int J Environ Sci Technol 11(1):191–198
28. Weber J, Kurková R, Klánová J, Klán P, Halsall CJ (2009) Photolytic degradation of methyl-parathion and fenitrothion in ice and water: implications for cold environments. Environ Pollut 157(12):3308–3313
29. Yao Y, Lenhoff AM (2006) Pore size distributions of ion exchangers and relation to protein binding capacity. J Chromatogr A 1126(1–2):107–119
30. Zakhama S, Dhaouadi H, M'Henni F (2011) Nonlinear modelisation of heavy metal removal from aqueous solution using *Ulva lactuca* algae. Bioressour Technol 102:786–796
31. Zheng L, Pi F, Wang Y, Xu H, Zhang Y, Sun X (2016) Photocatalytic degradation of acephate, omethoate, and methyl parathion by $Fe_3O_4@SiO_2@mTiO_2$ nanomicrospheres. J Hazard Mater 315:11–22

Performance Evaluation of Sewer Networks Lateral Ventilation Using Series of Inclined Jets in Preventing Obstruction Phenomenon

Hella Adouni, Yoldoss Chouari, Hervé Bournot, Wassim Kriaa, and Hatem Mhiri

Abstract Channel blockage is a very widespread phenomenon in sewer systems. The following paper presents a study of the optimal number of injectors in lateral ventilation using inclined jets in order to overcome the blockage phenomenon in ventilated confined channels. Numerical simulations were performed to evaluate the effect of a series of lateral jets on the air–water flow dynamics and to define the optimal distance between each pair of injectors. A strong interaction was found between the lateral air injection and the air–water interface inside the channel. The proposed ventilation technique disrupts the formation of a fully developed slug. It was also found that lateral injection with the described optimal distance between each pair of injectors completely removes channel blockage.

Keywords Obstruction channel phenomenon · Side by side jets · Air–water slug flow · Sewer system ventilation

1 Introduction

The monitoring of the proper disposal of rainwater and wastewater after treatment in ejection points such as the sea is essential for the protection of the environment. Indeed, the evacuation networks of these waters are generally confined pipelines featuring a two-phase air–water flow separated by a deformed interface due to a wave field under certain conditions. During intense rainfall events, the discharge capacities of the stormwater systems are often exceeded and the flow changes from a smooth

H. Adouni (✉) · Y. Chouari · W. Kriaa · H. Mhiri
LTTPI, National Engineering School of Monastir, University of Monastir, Monastir, Tunisia
e-mail: hella.adouni2@gmail.com

H. Bournot
CNRS, IUSTI, Aix Marseille University, Marseille, France

R. Khiari and M. Jawaid (eds.), *Proceedings of the 3rd International Congress of Applied Chemistry & Environment (ICACE-3)*, Springer Proceedings in Materials 23, https://doi.org/10.1007/978-981-99-1968-0_12

stratified flow to a slug flow blocking the channel. At this point, hydraulic problems occur that threaten the infrastructure of the sewerage systems. These include geysers, water hammer, leaks, broken pipes, manhole overflows leading to flooding, violent expulsion of manhole lids, air entrapment and blockage of pipes leading to the escape of undesirable and toxic odors threatening human health and polluting the air. Considering all these problems, the study of two phase flow dynamics in sewerage systems has turned out to be a very relevant issue. Akhlaghi et al. [1] conducted an experimental and numerical study aimed at better understanding the characteristics of a slug flow. The experiments were performed in a horizontal pipe of diameter equal to 44 mm. The VOF method was used for the simulation of the flow. The study showed that the effect of air superficial velocity is more important in terms of the slug initiation position, on the other hand the slug frequency is more sensitive to the liquid superficial velocity than that of air. Akhlaghi et al. [2] numerically studied plug flow in horizontal pipe using VOF model. The superficial velocity of water was fixed at 1 m/s and that of air varied from 0.16 to 3 m/s. For a fixed water velocity, when the air velocity increases, the liquid retention decreases, revealing that the gas phase section is indeed dominant. Nasrfard et al. [3] simulated the intermittent two-phase flow in a horizontal pipe of internal diameter D = 54 mm for low gas superficial velocities ($V_{sg} < 1$ m/s). The results showed that the slugs move closer to the entrance as the gas superficial velocity increases and that the position of their occurrence is independent of the liquid superficial velocity. Conte et al. [4] conducted an experimental study intended to better understand the characteristics of a slugged flow. The experiments were performed in a horizontal pipe of diameter D = 26 mm. The ranges of the superficial velocities of the gas and the liquid were set respectively to (0.2–1 m/s) and (0.3–0.7 m/s). The study showed that there is a strong dependence between the superficial velocity of the liquid and the initiation position of the slug. Adibi et al. [5] conducted an experimental study on the effect of liquid retention (liquid holdup) on slug flow for different superficial velocities in a horizontal pipe with a cross section of 5 cm × 10 cm. The results revealed that the created holdup tends to cause large pressure fluctuations when the fraction of the cross-sectional channel area that is occupied by the gas phase (void fraction) is not equal to the one occupied by the liquid phase. Most of the existing studies in the literature have been interested in investigating the internal structure of a two-phase slug flow in a confined channel. A minority of researchers has focused on finding a solution to the problem of slug flow to prevent overflows and pipes obstruction in sewage systems and, therefore, eliminates the problem of toxic gas backflow. Adouni et al. [6] studied the dynamics of slug flow in a horizontal channel by investigating the effect of superficial air velocity on the flow pattern. Numerical correlations capable of predicting the time and position of slugs' occurrence were presented. Such results are efficient in anticipating slug flow occurrence and thus avoiding sewer systems obstruction. Adouni et al. [7] proposes a solution that completely avoids the occurrence of slug flow inside confined channels. The proposed solution is a New Ventilation Technique (NTV) inside the confined channels. It consists of the induction of air flow laterally by external, parietal jets directed towards the interior of the channel. The results allowed determining the optimal axial lateral position and the orientation of the jets to ensure the elimination of the blockage problem in confined channels with two-phase flow. This paper is inspired by this particular topic and supports the validity of the solution presented by Adouni et al. [7] by means of a parametric study of the optimal number of injectors distributed on the side walls of the channel.

2 Material and Methods

2.1 Geometry Configuration

The simulated model is a rectangular horizontal channel with a height equal to 0.15 m and a width equal to 0.075 m, as the test bench of the IUSTI laboratory of the University of Aix Marseille where experiments were performed. Water is introduced from the channel water inlet. Air is introduced longitudinally from the channel inlet and laterally from the opposed inclined parietal jets as shown in Fig. 1.

2.2 Numerical Modeling

The two phase air–water flow is unsteady and turbulent. Air and water are assumed to be incompressible fluids. We considered the superficial tension effect between both introduced fluids. The equations governing the flow are the conservation of mass (1) and the conservation of momentum (2):

$$\frac{\partial \rho u_i}{\partial x_i} = 0 \tag{1}$$

$$\frac{\partial \rho u_i}{\partial t} + \frac{\partial \rho u_i u_j}{\partial x_j} = -\frac{\partial P}{\partial x_i} + \frac{\partial \tau_{ij}}{\partial x_j} - \rho g + f_{vol} \tag{2}$$

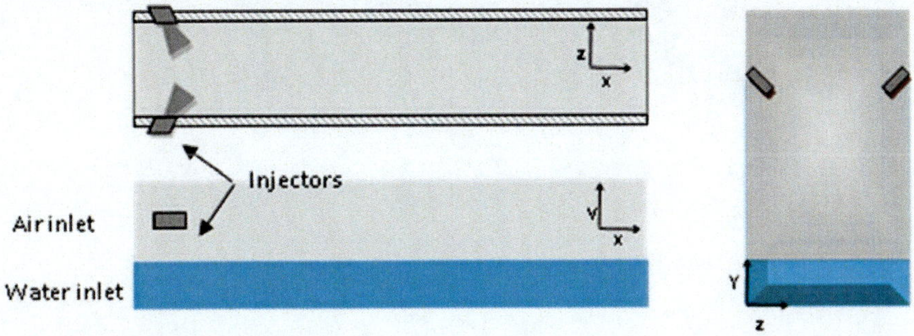

Fig. 1 Layout of the channel

Table 1 Initial conditions

$V^{injector}$ [m/s]	$V^{s}{}_{air}$ [m/s]	$V^{s}{}_{water}$ [m/s]
0.25	12.95	0.16

where ρ is the density of the fluid, t is time, u the velocity of the fluid, f_{vol} is the volumetric force due to superficial tension and τ the stress tensor.

$$\tau_{ij} = \mu\left(\frac{\partial u_i}{\partial x_j} + \frac{\partial u_j}{\partial x_i}\right) - \frac{2}{3}\mu\delta_{ij}\frac{\partial u_l}{\partial x_l} \quad (3)$$

where μ is the viscosity and δ is the Kronecker symbol.

The turbulence model is used to solve the governing equations in the "RNG" k-epsilon model. For the representation of the interface, the most appropriate model is the Volume of Fluid model. The initial conditions are resumed in Table 1. The boundary conditions are plotted in Fig. 2.

The numerical simulation is performed using the ANSYS Fluent 16.2 commercial code. The volume fraction discretization scheme used is the geo-reconstruction scheme. A second order upwind scheme is used for the discretization of the transport equations (turbulent kinetic energy equation and dissipation rate equation). The coupling of the pressure–velocity is performed with the PISO algorithm.

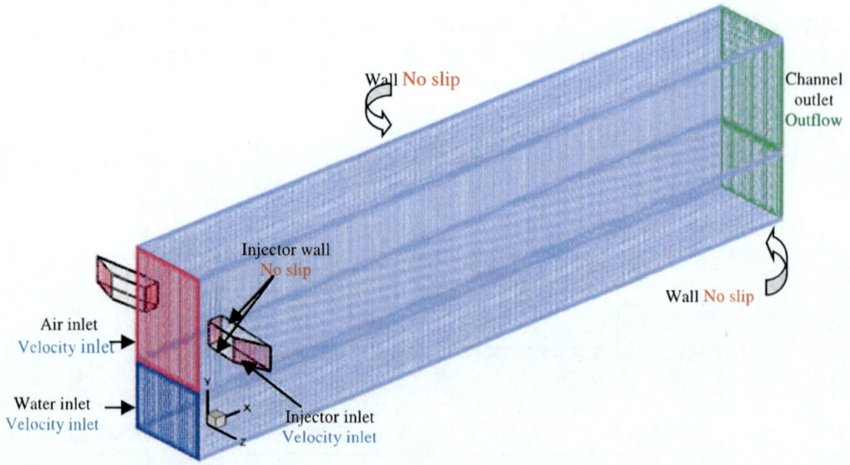

Fig. 2 Boundary conditions

3 Results and Discussions

Firstly, we propose to study the aspect of the flow in the case of n series of successive jets. Secondly, the optimization of the distance between the successive series of jets will be examined. And finally, the solution allowing the complete elimination of the slug flow will be presented.

3.1 *Study of the Efficiency of the NTV with "n" Series of Jets*

Figure 3 shows the air volume fraction contours of the configuration without injectors (a) and with n sets of injectors: a set of 2 injectors (b), 4 injectors (c), 6 injectors (d) and 8 injectors (e). The volume fraction contour without injectors shows the initiation of slug flow at a position $X_s/D = 1$. The injectors must be installed at a position lower than the position where the first wave is perceived ($X_s/D = 1$). Thus, the first two injectors are placed at $X_{inj}/D = 0.2$. The volume fraction contour with the first set of injectors (2 opposing injectors) shows the initiation of a slug flow at $X_s/D = 4$. Accordingly, the introduction of the second set of injectors was carried out at an axial position $X_{inj} < X_s$ which pushed the disturbances further downstream of the entrance. Following the same principle, we introduced the third and fourth sets of injectors. These contours showed that the introduction of n series of lateral injectors has a direct effect on the air–water flow dynamics inside the channel. It shows a change in the structure of the two-phase flow. The interface becomes very fragmented. It is covered by several rolling waves of large amplitude rather than by the development of a slug. The introduction of n series of lateral injectors has attenuated the flow by triggering large amplitude waves that are struggling to form a slug. These waves can then break up or merge with other large amplitude waves to form a pseudo-slug or a slug later. Thus, the obstruction of the channel is delayed.

3.2 *Study of the Optimal Distance Between the of "n" Ejects*

Figure 4 shows the evolution of the axial velocity along the channel depending on the number of injectors. According to the pink curve which represents the case of the first series of injectors (2 injectors), we notice that in a first time the velocity increases to reach a maximum value from which it decreases further downstream to a constant value.

The increase in speed is the result of the merger of the two opposing jets with the main air flow. This increase in velocity will push the interfacial instabilities downstream. Therefore, from the point where the velocity begins to decrease, it is pertinent to add another series of two jets to push the airflow that tends to create problems on the air–water interface.

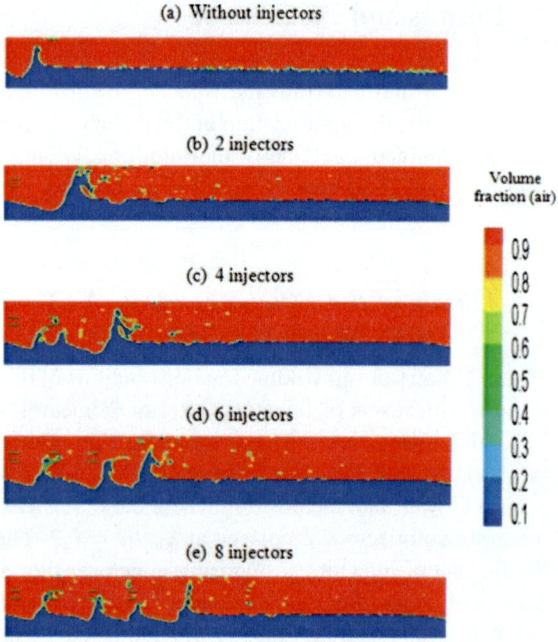

Fig. 3 Air volume fraction contours of the initial configuration (**a**) and the new configuration with n sets of injectors: a set of 2 injectors (**b**), 4 injectors (**c**), 6 injectors (**d**) and 8 injectors (**e**)

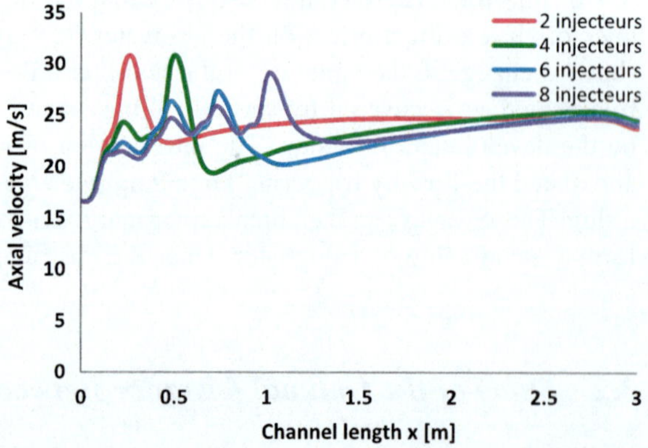

Fig. 4 Evolution of the axial velocity for (x, $y = 0.5$ Hair, $z = 0.5Z$) according to the number of injectors

The second series added, described by the green curve, shows a second zone where the velocity resumes its increase after its stagnation at the end of the first series of injectors. After each set of injectors, the velocity decreases and tends to stagnate until the next set of injectors where it will increase again. These curves allowed us to show the usefulness of installing injectors side by side along the channel and to determine the optimal distance between two successive series of injectors which is $X = 2.6D$.

Thus, 12 successive sets of jets were introduced to the configuration with a distance $X = 2.6D$ between each two sets, as shown in Fig. 5. This technique eliminated the

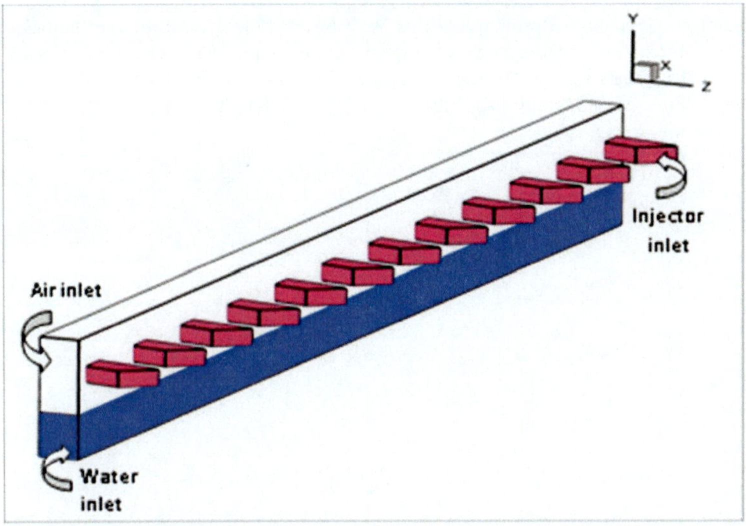

Fig. 5 Side by side lateral injectors

occurrence of slug flow in the channel and therefore solved the phenomenon of blockage in confined pipes.

4 Conclusion

The introduction of jets arranged side by side along the lateral walls of the channel with a distance between each two series of about 2.6*D* overcame the phenomenon of blockage in the confined channels.

References

1. Akhlaghi M, Taherkhani M, Nouri NM (2020) Study of intermittent flow characteristics experimentally and numerically in a horizontal pipeline. J Nat Gas Sci Eng 79. https://doi.org/10.1016/j.jngse.2020.103326
2. Akhlaghi M, Mohammadi V, Nouri NM, Taherkhani M, Karimi M (2019) Multi-fluid VoF model assessment to simulate the horizontal air–water intermittent flow. Chem Eng Res Des 152:48–59
3. Nasrfard N, Rahimzadeh H, Ahmadpour A, Amani E (2019) Simulation of intermittent flow development in a horizontal pipe. J Fluids Eng Trans ASME 141(12):1–10
4. Conte MG, Hegde GA, Marco J, Sum AK, Morales REM (2017) Characterization of slug initiation for horizontal air–water two-phase flow. Exp Therm Fluid Sci 87:80–92
5. Adibi AP, Ansari MR, Jafari S, Habibpour B, Salimi E (2014) Slug initiation evaluation in long horizontal channels experimentally. In: 2014 International conference on fluid mechanics and applications (ICFMA)

6. Adouni H, Chouari Y, Bournot H, Kriaa W, Mhiri H (2022) A novel ventilation method to prevent obstruction phenomenon within sewer networks. Int J Heat Mass Transf 184
7. Adouni H, Chouari Y, Kriaa W, Bournot H (2021) The effect of air velocity on slugs in a confined channel. Flow Meas Instrum J 81. https://doi.org/10.1016/j.flowmeasinst.2021.102013

Tear Strength Properties of Recycled Denim Fabrics Under Mixed Washing Using Smart Foam Technology

Cheriaa Rim and Ben Marzoug Imed

Abstract Seven denim fabrics were produced with yarns obtained from cotton fibers (CO), recycled polyester (r-PET), recycled cotton fibers (r-CO), and recycled Tencell fibers (r-Tencell). A mixed washing process using cellulase enzyme and synthetic stones was applied on these fabrics using Smart foam technology. We established using MINITAB® 17.1.0 statistical software a mixed-level factorial design of experiments to study the effects of 2 categorical factors: recycled fiber percentage and enzyme type and 3 continuous factors: treatment time, enzyme dosage and synthetic stone percentage. The output or the response is the tear strength loss of treated fabrics. The main goal of this work is to keep fabric tear strength within tolerable limits (>2500 cN). Critical and optimal processes were obtained using response surface methodology (RSM) and desirability functions. Correspondent tear strength losses were discussed.

Keywords Recycled denim fabrics · Tear strength properties · Mixed washing · Smart foam technology

1 Introduction

Ecological procedure engaged in denim production simplifies innovative steps in denim washing named waterless washing: Lazer denim finishing, Ozone denim finishing, E-flow technology… Enzymes come from renewable energy and using auxiliary of corrosive chemicals for chemical washing in one-bath leads to enzymatic fading [13]. To address the environmental concerns, some other finishing techniques have been introduced as an alternative to the conventional wet processing. Smart foam, an innovative patent-pending technology, has the objective of revolutionizing garment finishing. The system is intended to minimize natural resource consumption: water and energy, making sustainable garment production easy and accessible (3

C. Rim (✉) · B. M. Imed
Laboratory of Textile Engineering (LR11ES42), University of Monastir, Monastir, Tunisia
e-mail: cheriaar@yahoo.fr

R. Khiari and M. Jawaid (eds.), *Proceedings of the 3rd International Congress of Applied Chemistry & Environment (ICACE-3)*, Springer Proceedings in Materials 23, https://doi.org/10.1007/978-981-99-1968-0_13

times faster to load chemicals in the washing machine). It is a result of a collaboration of Garmon, Itaclab and Mactec, three leading companies across the garment finishing industry. The main benefit of smart foam is the great reduction of water and energy consumption: With smart foam the chemical carrier is not water. Compared to traditional washing processes, the application of chemicals through smart foam allows savings of up to 80% of water. All the treatments are performed at room temperature, therefore reducing the energy required. Any type of washing machine can be utilized with the smart foam equipment and process [5].

Garment wastes are an extra important problematic [3]. Moreover, per capita fiber consumption in the world has been growing steadily depending on the increase in population and income level. The solution found or these problems is recycling processes i.e. mechanical, thermal, chemical and biological processes. Compared to virgin fibers, the fiber length of the recycled staple fibers becomes significantly lower. The negative structures of yarns and fabrics produced from recycled fibers have been studied in previous studies [7, 8, 14]. It was known from these investigations that recycled yarn has lower tensile strength, higher unevenness, and higher imperfection index of yarns values. There is a tendency to blend the virgin material to compensate the losses [4]. Working with higher yarn densities can hide the negative characteristics of recycled yarns [6]. Fabric appearance and hand feel will not be affected negatively by the help of the finishing process as long as denim fabrics produced from recycled yarns have acceptable physical and mechanical properties [1, 6]. Nevertheless, it is pointed out in the literature that washing process can deteriorate fabric mechanical properties, especially tear and tensile strength of the products [10, 11]. Finally, the negative effects of recycled fibers can be re-emerged on the fabric in more abrasive washing processes in spite of the working with higher yarn weight in denim sector [12].

2 Experimental

2.1 Materials

7 indigo dyed denim fabrics which mass area ranged from 314 to 462 g/m^2 were used in this study. Fabrics will be referred as Fabric a–g. The same washed denim fabrics were used previously to study tear strength properties using E-Flow technology [2] (Table 1).

Table 1 Fabric composition

Category 1		Category 2		Category 3		
A	b	c	d	e	f	g
90% Co/ **10% r-Co**	69% Co/ **30% r-Co/** 1% Elas.[a]	13% Co/ **50% r-Co/ 36% r-Tencell/** 1% Elas.	16% Co/ **36% r-Co/ 48% r-Tencell**	77% Co/ **7% r-Co/ 14% r-PES/** 2% Elas.	67% Co/ **16% r-Co/ 16% r-PES/** 1% Elas.	66% Co / **12% r-Co/ 20% r-PES/** 2% Elas.

[a] Elas.: Elasthane, Co: Cotton

Table 2 Mixed washing recipe

Action	Liquor ration	Time, (min)	Temperature, (°C)	Chemical	Dosage
Rinsing	1/6	2	30	Water	–
Mixed washing	**–**	**30-60**[a]	**30**	**Enzyme** **Synthetic stones** Foaming agent anti redeposant agent	**1–1.5 - 2%**[a] **0–100–200%** 20% 10%
Rinsing	1/6	2	30	Water	–
Enzyme deactivation	1/6	5	40	Na_2CO_3	1.5%
Rinsing	1/6	2	30	Water	–
Drain	–	2	–	–	–
Drying	–	40	60	–	–

[a] See Table 3

2.2 Methods

Table 2 shows the mixed washing recipe. Synthetic stone washing was combined with different neutral cellulase enzyme to reduce the usage of pumice stone. Smart foam technology was used and 3 commercial cellulase enzymes DS (A); NCS (B), NEF (C) were purchased.

2.3 Testing and Analysis

Tear strength was evaluated according to ASTM D 1424-1996: Standard Test Method for tearing strength of fabrics by the falling-pendulum type (Elmendorf) apparatus.

Table 3 Factors and levels related to enzymatic stone treatment for each denim category

Coded input		Uncoded input parameters	Levels		
			1	**2**	**3**
Categorical factors	x(1)	Recycled fiber percentage, %	P1[a]	P2[a]	P3[a]
	x(2)	Enzyme ID	A	B	C
Continuous factors	x(3)	Treatment time, min	30	60	–
	x(4)	Enzyme, %	1	1.5	2
	x(5)	Pumice stone, %	0	100	200

[a] See details in Table 1

2.4 *Experimental Design*

Table 3 shows input parameters and corresponding levels used in a mixed level factorial design of experiments.

3 Results and Discussion

Table 4 resumes ANOVA analysis used to determine the significance of each variable and interactions on the response.

Table 4 shows that recycled fibers %; treatment time; enzyme % and enzyme type had statistically significant effects on tear strength loss with a p-value <0.05 in almost categories. Of the all-two-way interactions, the relationship between r-fibers % and the tear strength loss is highly dependent on the enzyme type, synthetic stones % and treatment time. The R-sq values are acceptable (>89%) expect for category 2 (27.05%), an indication that the models explained more than 80% of the variance in tear strength loss (%) after treatment.

Table 5 shows that the increase of recycled fibers percentage contributes to the fabric tear strength loss to a greater extent. The abrasive action of synthetic stones and enzyme % have a non-significant effect on the fabric tear strength loss. Treatment time enhances tear strength loss of fabrics from category 1, 2 and 3. Selected enzymes had different effects on tested fabrics which is explained by their specific relative bio stoning activity. However, tear strength increases under mixed washing in some cases which can be due to the increased mobility of yarns within the fabric structure and the grouped yarns against tear [9].

Table 6 presented some regression models relative to denims containing r-cotton needed for optimization, and each one presents the relationship between tear strength loss and significant variables and interactions when the recycled fibers % and the enzyme type were on hold.

Many optimized processes independent of the fabric were obtained using desirability functions to minimize tear strength loss in warp and in weft direction.

Table 4 Summary of variance analysis results

Fabric category	Direction	Significant factors (p-value < 0.05)	Significant interactions (p-value < 0.05)	R-sq, (%)
1	Warp	• % r-cotton • Time • Enzyme type	• % r-cotton * enzyme type • time * enzyme % • time * synthetic stones % • time * enzyme type • time * % r-cotton • enzyme % * synthetic stones % • enzyme type * % r-coton	95.61
	Weft	• % r-cotton • Time • Enzyme type	• % r-coton * synthetic stones % • % r- cotton * enzyme type • Time * synthetic stones %	89.26
2	Warp	• % r-Tencell	–	27.05
	Weft	• % r-Tencell • Time • Enzyme % • Enzyme type	• % r-Tencell * enzyme type	89.52
3	Warp	• % r-PES • Time • Enzyme type	• % r-PES * time • synthetic stones % * enzyme type	98.83
	Weft	• % r-PES • Time • Enzyme type • Synthetic stones %	• % r-PES * synthetic stones % • % r-PES * time	93.19

Table 5 Summary of main effects plots

Category	Response: Tear strength loss, %	Factors				
		Recycled fibers, %	Enzyme type	Time, min	Enzyme, %	Pumice stone %
1 (r-Co)	Warp	++	–.–	+	Ø. Ø	Ø. Ø
	Weft	++	–.–	+	Ø. Ø	Ø. Ø
2 (r-Tencell)	Warp	++	–.Ø	+	+.–	++.–
	Weft	–	Ø.–	+	Ø. Ø	Ø. Ø
3 (r-PES)	Warp	+.++	–.Ø	+	Ø. Ø	Ø. Ø
	Weft	–.++	–.Ø	+	Ø. Ø	+.+

– No significant effect
++ Highly positive significant effect, + positive significant effect, no significant effect,
–.– Highly negative significant effect, –negative significant effect
Ø no significant effect

Table 6 Regression Models

Enzyme ID x(2)	R-cotton, % x(1)	Tear strength loss, %
A	10	−0.09 + 0.0417 x(3) – 2.61 x(5) + 0.0785 x(3)* x(5)
	30	21.33 + 0.0417 x(3) – 2.61 x(5) + 0.0785 x(3)* x(5)
B	10	−6.26 + 0.0417 x(3) – 2.61 x(5) + 0.0785 x(3)* x(5)
	30	19.49 + 0.0417 x(3) – 2.61 x(5) + 0.0785 x(3)* x(5)
C	10	−10.07 + 0.0417 x(3) – 2.61 x(5) + 0.0785 x(3)* x(5)
	30	18.10 + 0.0417 x(3) – 2.61 x(5) + 0.0785 x(3)* x(5)

Howover, a standard critical process is obtained: 60 min treatement, 2% enzyme, and 200% synthethic stones. The tear strengh loss under critical process in each denim category is a function of recyled fibers %, fabric composition, and fabric structure (yarn density, yarn strength…).

4 Conclusions

The tensile strength of the fabric has been reduced due to the mixed washing. The washing parameters had influenced the tear strength properties of recycled denim garments, especially denim fabrics made from r-Co and r-Tencell. Despite the tear strength loss reaches 60%, particularly in warp direction (denim fabric with r-Tencell), the tear strength value stays industrially acceptable (>2500 cN).

Bio stone washing time is the main influencing parameter on the tear strength of recycled denim garments. Recycled denim fabrics using r-Co, r-Tencell and r-PET fibers can show similar fabric tear strength loss properties to original denim. As found in our previous work [2], there isn't any clear benefit of r-PET fibers in the prevention of tear strength loss after mixed washing. To conclude about the best enzyme to use, it is required in future work to have a look on the desired effects: hand feel, redeposition, color change… Results can be also compared to our previous findings obtained using E-Flow technology.

References

1. Aronsson J, Persson A (2020) Tearing of post-consumer cotton T-shirts and jeans of varying degree of wear. J Eng Fibers Fabr 15
2. Cheriaa R, Ben Marzoug I (2022) The effect of recycled fibers on the ecological washing performance of denim fabrics, Ch. 15. Springer Science and Business Media LLC
3. Dobilaite V, Mileriene G, Juciene M, Saceviciene V (2017) Investigation of current state of pre-consumer textile waste generated at Lithuanian enterprises. Int J Cloth Sci Technol 29:491–550

4. Esi B, Bayka PD (2020) Investigation of tensile strength and elongation properties of chenille upholstery fabrics including recycling polyester yarns. J Eng Fibers Fabr 15:1–10
5. Garmon (2022) https://www.garmonchemicals.com/en/textile-chemicals/product-lines/smart-foam
6. Gulich B (2006) Development of products made of reclaimed fibres. In: Wang Y (ed) Recycling in textiles. Woodhead Publishing, Cambridge
7. Halimi MT, Hassen MB, Azzouz B, Sakli F (2007) Effect of cotton waste and spinning parameters on rotor yarn quality. J Text Inst 98:437–442
8. Hasani H, Semnani D, Tabatabei S (2010) Determining the optimum spinning conditions to produce the rotor yarns from cotton wastes. Ind Text 61:59–64
9. Hu J (2004) Modelling drape deformation of woven fabrics and garments—theory. Structure and mechanics of woven fabrics
10. Kan CW (2014) Effect of enzyme washing on the tensile property of denim fabric. Adv Mater Res 933:175–178
11. Mezarcıöz S, Toksöz M (2014) Investıgatıon of Effect of dpecıal washıng processes on denım fabrıcs' propertıes. Tekstil Ve Konfeksiyon 24:68–95
12. Rahman Khan MM, Mondal IMH, Uddin MZ (2012) Sustainable washing for denim garments by enzymatic treatment. J Chem Eng, IEB 27:27–31
13. Shibly MAH, Hoque MM, Miah S (2021) Development of eco-friendly denim fabric washing by natural resources. Int J Text Sci 10:1–6
14. Telli A, Özdil N (2015) Effect of recycled PET fibers on the performance properties of knitted fabrics. J Eng Fibers Fabr 10:47–60

Sustainable Dyeing Process of Alfa Handicrafts from Kasserine Using Natural Dyes

Samar Mansouri, Noureddine Baaka, and Hatem Dhaouadi

Abstract Alfa is an herbaceous perennial plant growing in the southern Mediterranean, particularly in Spain and the Maghreb. In Tunisia, esparto is found in the central, western, and southern regions, such as Kasserine, Gafsa, and Sidi BouZid. Alfa constitutes the raw material of multiple traditional handicrafts such as the couffins and scourtins used for the extraction of olive oil, mats, and carpets. The alfa is also traditionally used for the manufacture of baskets and bissacs for transport, grain silos, or even sandals and ropes. Currently, alfa fibers are dyed using only synthetic dyes. This work studies the development of a sustainable dyeing process of Alfa handicraft products using natural dyes instead of synthetic dyes. Different colors have been obtained from three plant sources. The dyeing process performances were evaluated by measuring CIELab parameters and colour yield values of dyed fabrics. Authors, also, studied the effect of mordant type with different mordanting products (ferrous sulphate, and alum) on dyeing quality.

Keywords Natural dyeing · Handicrafts · Alfa (*Stipa tenacissima*)

S. Mansouri (✉) · N. Baaka · H. Dhaouadi
Faculty of Sciences, Research Laboratory of Environmental Chemistry and Cleaner Process (LCE2P-LR21ES04), University of Monastir, 5000 Monastir, Tunisia
e-mail: samar20481@yahoo.fr

S. Mansouri
Higher Institute of Arts and Crafts of Kasserine, Kasserine, Tunisia

N. Baaka
Higher Institute of Fashion of Monastir, 5019 Monastir, Tunisia

R. Khiari and M. Jawaid (eds.), *Proceedings of the 3rd International Congress of Applied Chemistry & Environment (ICACE–3)*, Springer Proceedings in Materials 23, https://doi.org/10.1007/978-981-99-1968-0_14

1 Introduction

1.1 Alfa

Alf is an herbaceous perennial plant which grows in the South of the Mediterranean, in particular in Spain and North Africa [1]. This wild species is resistant to dry and sunny climates, while its numerous roots allow it to grow in relatively poor soils. Alfa grows in dense clumps up to 1 m high. Its robust and leathery leaves let appear flowers in spring.

Alfa is considered as the oldest used material. First of all, it was used as an animal's food due to its availability. Then, due to its morphological properties (linear form), Alfa can be transformed to linear, bi-dimensional, and three-dimensional fabrics. It has also always been used by the Bedouins of the region in their movements and their halts for the manufacture of the ropes serving to fix the tents. After that braided ropes were used to produce carpets. The famous mats produced after weaving Alfa leaves was a good solution for Man to resolve the problem of his mattress. Now, these fabrics are certified to be comfortable for orthopedic uses.

This plant is also used to make several other three-dimensional objects for the conservation of harvests and the transport of foodstuffs such as fruits and vegetables purchased or collected while on the move [2].

All mentioned uses are produced manual; they are handmade. In industry, alfa fibers are transformed into pulp and are used to manufacture various papers, but also bank bills and industrial filters [3, 4]. For instance, in Tunisia, the production of Alfa pulp exceeds 30,000 tons per year Despite of the importance of this utilization (papermaking) the handmade of Alfa still occupies an important position, especially when consumers are looking for using biological fabrics in all domains. In the same context of using natural materials it is very important to add natural products in finishing process fabrics such the dyeing process [5–7].

1.2 Natural Dyes

Man has put colors in his life since the beginning of his adventure. He first used pigments from "colored earths", then those from plant and animal fibers [8]. Whether using raw, refined, or synthetic pigments, humans quickly created palettes that millennia have failed to erase [9]. There are two types of dyes namely natural dyes which are pigments from vegetables, fruits, seeds, or spices and synthetic dyes which are totally from chemical synthesis [10].

Synthetic dyes have the advantage over natural dyes of having a longer shelf life and giving more intense colors. They are therefore used in smaller quantities and are less expensive than natural dyes that are sensitive to light, oxygen, or bacterial action [11].

However, natural dyes are considered non-toxic, non-allergic, and biodegradable [12–15].

In view of the environmental problems caused by the use of synthetic dyes [16], the textile industry is faced with the challenge of replacing synthetic dyes with natural dyes [17].

1.3 Objective of this Study

The purpose of this study is to develop a sustainable dyeing process of alfa handicrafts using three natural dyes namely which are turmeric rhizomes, madder roots, and henna leaves. The effect of various metal salts (ferrous sulphate, and alum) as mordants on color parameters of dyed samples were comparatively evaluated. CIELab parameters (L*, a*, b*) and the colour yield (K/S) were used to evaluate the shades of the dyed samples.

2 Experimental

2.1 Materials

The Alfa leaves used during this work were collected from the region of Kasserine, which is located in west-central Tunisia, in August 2020.

Madder barks (Fig. 1b) and henna leaves (Fig. 1c) were both collected in December 2020, from the region of Gabes, which is located in southeastern Tunisia. Turmeric rhizomes (Fig. 1a) were bought on the local market. Photographs of these three materials are presented in Fig. 1a–c. All plant materials were dried in the shade. Then the samples were ground and used for dyeing.

2.2 Dyes Extraction Method

The extraction process of dyes from each material is as follows: 5 g of each sample was mixed with 100 mL of distilled water. The mixture was heated, with stirring, for 1 h at a temperature of 90 °C.

The obtained colored solutions were filtered using an ordinary cotton cloth and used as dye bath.

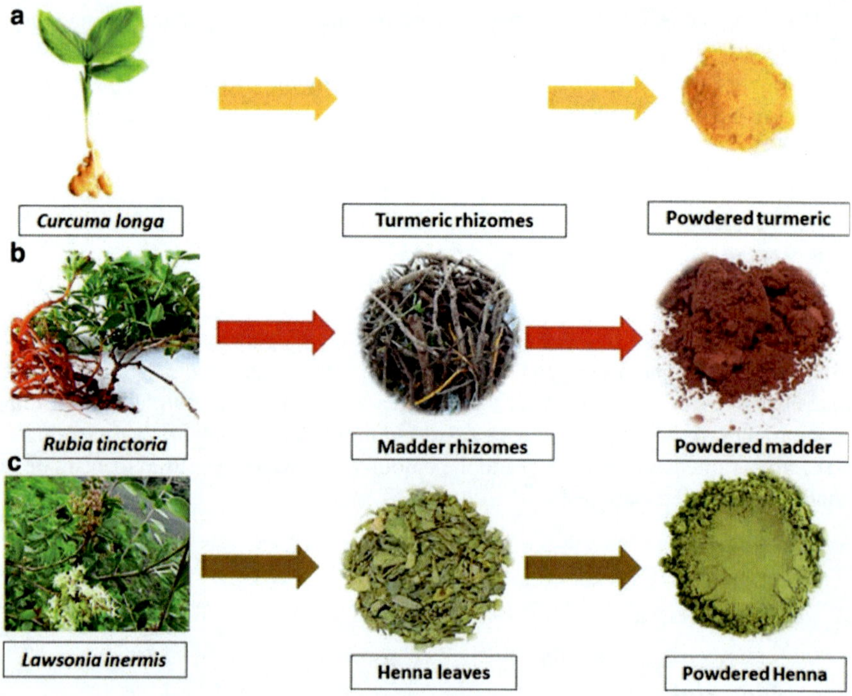

Fig. 1 **a** *Curcuma longa*, **b** *Rubia tinctorum* L. and **c** *Lawsonia inermis*

2.3 Mordanting and Dyeing Method

During this study, the pre-mordanting method was used for mordanting the alfa leaves using an amount of with 5% (ww) ferrous sulfate ($FeSO_4 \cdot 7H_2O$) and 10% (ww) alum ($Al_2K_2(SO_4)_4 \cdot 24H_2O$).

This method consists of immersing the alfa leaves in mordant solutions at a temperature of 30 °C with a liquor ratio of 40:1. Then the temperature was gradually increased to 60 °C at a rate of 2 °C per minute and maintained for 30 min. Then, the mordanted samples were thoroughly rinsed.

The dyeing without mordant was performed in a 40:1 liquor ratio under the following experimental conditions:

- 90 °C temperature,
- a duration of 60 min,
- a pH value of 7.

Finally, the samples were rinsed with water, several times. Then they were dried in the dark.

3 Results and Discussions

3.1 Color Shades of Dyed Alfa Leaves

The obtained dyed samples are given in Table 1 with different color shades depending on used substance and followed process.

For all the dyeing material we remark the same behavior of alfa leaves. In that, obtained shades without mordant substances were accentuated to be darker when

Table 1 Colour shades of dyed alfa leaves

Sample	Shades
Madder	
Unmordanted	
5% $FeSO_4$	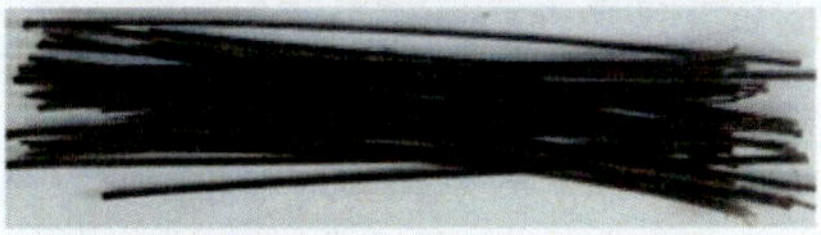
10% alum	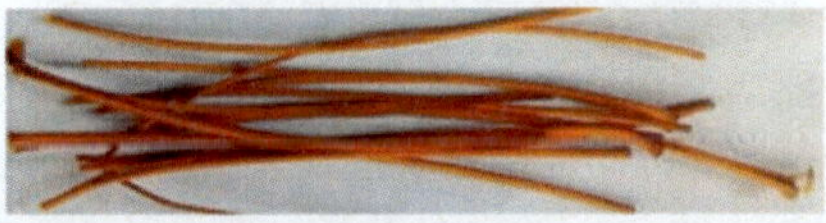
Henna	
Unmordanted	
5% $FeSO_4$	
10% alum	

(continued)

Table 1 (continued)

Sample	Shades
Turmeric	
Unmordanted	
5% $FeSO_4$	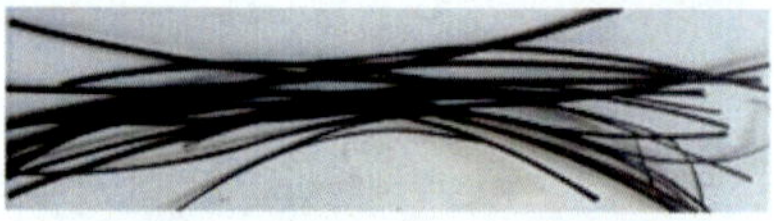
10% alum	

using ferrous sulfate as mordant. But after using alum, obtained color shades are brighter.

3.2 *Colorimetric Parameters of Dyed Samples*

It is important to declare that higher colour strength values indicate higher dyeing yield. As a result, deeper colors can be obtained. Colorimetric properties of un-mordanted and mordanted dyed alfa leaves were given in Table 2.

It is noted that the best values are obtained by using ferrous sulphate as a mordant. Thus, ferrous sulphate provided dark hues. This may be due to the ability of iron to form coordination complexes with the dyes used [6]. This metal can form a ternary complex at one site with the alfa leaves and at the other site with the dye substances. As a result it is the most important role of the mordant in the dyeing operation.

In contrary, alum leads to K/S value inferior to that of the un-mordanted ones only when using Turmeric for dyeing Alfa leaves. Also 10% of Alum added to the dyeing bath, increased generally the a* and b* coordinates. Thus, comparing to results obtained with ferrous sulphate, the shades became more red, yellow, and bright.

Table 2 Colorimetric properties of dyed samples

Sample	K/S	L*	a*	b*
Madder				
Unmordanted	9.20	55.03	18.19	44.29
5% $FeSO_4$	11.62	29.26	7.98	17.07
10% alum	6.89	38.92	21.26	40.02
Henna				
Unmordanted	10.25	31.49	20.23	44.22
5% $FeSO_4$	12.89	27.49	21.12	39.35
10% alum	10.02	34.98	22.52	45.25
Turmeric				
Unmordanted	7.52	70.80	5.49	64.91
5% $FeSO_4$	11.36	59.53	11.71	55.61
10% alum	8.41	69.25	6.95	73.70

3.3 Presentation of Prototype Products

After dyeing stage, the craftswomen use dyed alfa leaves to produce different products and they affirm that many problems encountered when using chemical dyes were avoided such as causing cough and coloring hands during work (Table 3).

Table 3 Presentation of two prototype products

Products	Descriptions	Pictures
Handicraft: under dish	Designation: under dish Function: table decoration and protection Material: alfa leaves dyed with madder Realization: was produced using spiral stitch technique	
Handicraft: basket	Designation: basket Function: put in bread or fruit Material: alfa leaves dyed with henna Realization: produced using spiral stitch technique	

4 Conclusion

The side effects of carcinogens have made people look to green and sustainable alternatives. Once woven for purely functional use, alfa is now also used for aesthetic and decorative purposes. Its resistant leaves of a beautiful natural color are indeed a raw material of choice in the manufacture of authentic and original decorative objects. In order to give added value to these leaves, the handicrafts are dyed with sustainable natural dyes extracted from three plants instead of synthetic dyes. It is found that a variety of colors of various shades and tones are obtained by changing the type of the mordant.

References

1. Akchiche O, Boureghda MK (2007) Esparto grass (*Stipa tenacissima* L.), raw material of papermaking. First part. ХИМИЯРАСТИТЕЛЬНО О СІРЬЯ. 4:25–30
2. M'Hamdi M, Anderson PC (2013) Approche ethnoarchéologique d'outils et techniques de moisson de l'alfa (Stipa tenacissima) dans la région des Hautes Steppes en Tunisie: Première-sobservations et analyses. An interdisciplinary focus on plant workingtools, APDCA, pp 75–88
3. Nadji H, Brochier Salon MC, Bruzzèse C, Benaboura A, Belgacem MN (2006) Chemical composition and pulp properties of Alfa (*Stipa tenacissima*). Cellul Chem Technol 40(1–2):45–52
4. Marrakchi Z, Khiari R, Oueslati H, Mauret E, Mhenni F (2011) Pulping and papermaking properties of Tunisian Alfa stems (*Stipa tenacissima*)—effects of refining process. Ind Crop Prod 34:1572–1582
5. Ben Ticha M, Meksi N, Drira N, Kechida M, Mhenni MF (2013) A promising route to dye cotton by indigo with an ecological exhaustion process: a dyeing process optimization based on a response surface methodology. Ind Crop Prod 46:350–358
6. Haddar W, Ben Ticha M, Guesmi A, Khoffi F, Durand B (2014) A novel approach for a natural dyeing process of cotton fabric with *Hibiscus mutabilis* (Gulzuba): process development and optimization using statistical analysis. J Clean Prod 68:114–120
7. Baaka N, Mahfoudhi A, Haddar W, Mhenni MF, Mighri Z (2016) Green dyeing process of modified cotton fibers using natural dyes extracted from *Tamarix aphylla* (L.) Karst. Leaves. Nat Prod Res 31(1):1655–1662. https://doi.org/10.1080/14786419.2017.1285303
8. Ben Mansour H, Boughzala O, Dridi D, Barillier D, Chekir-Ghedira L, Mosrati R (2011) Les colorants textiles sources de contamination de l'eau: CRIBLAGE de la toxicité et des méthodes de traitement. Revue des sciences de l'eau. J Water Sci 24(3):209–238. https://doi.org/10.7202/1006453
9. Akakpo E, Badoussi ME, Gnacadja CK, Houngbo H, Dossou A, Azokpota P (2019) Le rocouyer (Bixa orellana), une source de biocolorant pour les industries alimentaires: revue analytique. Int J Biol Chem Sci 13(4):2332–2351
10. Baaka N, Khiari R (2023) A new approach for studying the dyeability of date palm residues fabric with sustainable natural dyes. In: Jawaid M, Midani M, Khiari R (eds) Proceedings of 2nd world conference on byproducts of palms and their applications. ByPalma 2020. Springer proceedings in materials, vol 19. Springer, Singapore. https://doi.org/10.1007/978-981-19-6195-3_15

11. Ben Ticha M, Baaka N, Mabrouk S, Boudokhane C, Dhaouadi H (2022) A sustainable application of turnsole (dyer's litmus) extract to dye protein fibers: development of a microwave-assisted dye extraction process. J Nat Fibers 19(9):3571–3584. https://doi.org/10.1080/15440478.2022.2072442
12. Bao J, Cai YM, Sun M, Wang G, Corke H (2005) Anthocyanins, flavonols, and free radical scavenging activity of Chinese bayberry (*Myrica rubra*) extracts and their color properties and stability. J Agric Food Chem 53:2327–2332
13. Sinha K, Saha PD, Datta S (2012) Response surface optimization and artificial neural network modeling of microwave assisted natural dye extraction from pomegranate rind. Ind Crop Prod 37:408–414
14. Sivakumar V, Vijaeeswarri J, Anna JL (2011) Effective natural dye extraction from different plant materials using ultrasound. Ind Crop Prod 33:116–122
15. Sinha K, Saha PD, Datta S (2012) b) Extraction of natural dye from petals of Flame of forest (*Butea monosperma*) flower: process optimization using response surface methodology (RSM). Dyes Pigm 94:212–216
16. Magda TR (2003) Going back to natural red colors (red sorghum, cochineal, and Monascus). Drink Technol Mark 7:1433–1594
17. Yılmaz F (2020) Investigating the usage of eucalyptus leaves in antibacterial finishing of textiles against Gram-positive and Gram-negative bacteria. J Text Inst 112(2):341–345. https://doi.org/10.1080/00405000.2020.1753394

Experimental Study of Some Parameters Influencing the Efficiency of a Desalination Membrane

Khaoula Hidouri, Ali Benhmidene, Hiba Akrout, and Bechir Chaouachi

Abstract The problem of water purification nowadays becomes everyone's first concern. For this reason our laboratory has carried out several research studies in the field of desalination and more precisely that of membranes. The different techniques for recycling polymer products which are the basis for the manufacture of different types of hydrophobic membranes that be used with the different types of solvents and non-solvents. In order to increase the porosity as a result of making the membranes more efficient. Recyclic membrane is done by the thermally induced phase separation (TIPS) method. The crystalline property of the polymer was studied by FTIR spectroscopy to know about functional group differences between the different membranes. All fabricated membrane it be characterized us using terms of thickness, contact angle, pore size, porosity, mechanical strength, and morphological analyzes using SEM, showed that a membrane was obtained which has a good porosity, a homogeneous pore size distribution, and a contact angle greater than 90 °C, i.e. hydrophobic. These membranes are therefore applied to vacuum membrane distillation to desalinate seawater or brackish water. By comparing the permeate flux results of prepared membranes with those of commercial membranes, we find almost similar results.

Keywords Membrane preparation · Membrane distillation (MD) · Thermally induced phase separation method (TIPS)

1 Introduction

So many natural resources it be used in our days to make water drinking. Several countries of the world due to the lack of sufficient natural water resources. Therefore, it was necessary to plan and create new methods to realize the availability of fresh water for

K. Hidouri (✉) · A. Benhmidene · H. Akrout · B. Chaouachi
Research Laboratory of Energy, Water, Environment and Processes, National Engineering School of Gabes, Omar Ibn El Khattab Avenue, 6029 Gabes, Tunisia
e-mail: khaoula2013@yahoo.fr

R. Khiari and M. Jawaid (eds.), *Proceedings of the 3rd International Congress of Applied Chemistry & Environment (ICACE–3)*, Springer Proceedings in Materials 23, https://doi.org/10.1007/978-981-99-1968-0_15

domestic, agricultural, and industrial consumption. Given the major availability of saline water in the earth, we will think of exploiting it in order to remedy the shortage of water and sulfur. In the year 2013, more than 17,000 desalination plants were distributed in more than 150 countries with a total capacity of about 80 million m^3/day. Figoli et al. [1] reported that 63.7% of the total capacity is produced by membrane processes and 34.2% by thermal processes. As the membranes present the main element of membrane processes which are the most used in the field of desalination. Composite membranes appeared in the early 1990s, they are characterized by an asymmetrical structure whose skin is much thinner than that of conventional non-composite membranes and by a superposition of several layers differentiated either by their chemical nature or by their physical state. They can be organic (superposition of different organic polymers), organo-mineral, or mineral (combination of carbon or alumina as support and metals such as zirconia, alumina, and titanium). There are two main types of filtration membrane geometries: flat membranes and membranes with cylindrical geometry (hollow fibers and tubular membranes). Flat membranes are in the form of a flat film whose mechanical strength is provided by a support (supported membranes). This film is usually thin (thickness less than a hundred micrometers) [2]. The dimensions of the tubular membranes are larger (dimensions in tens of millimeters). Hollow fibers have a very high compactness. For an equal filtration surface, an installation using flat membranes has a volume 16 times greater than a process using hollow fibres. The latter also have the advantage of having a longer service life, a degraded or broken fiber can be isolated from the rest of the filtration module which will continue to operate. The phase inversion method is the method chosen by manufacturers to manufacture the majority of flat membranes and hollow fibers because it allows the development of membranes that cover the entire filtration range. Phase inversion is the most widely used technique for the manufacture of polymer-based membranes. The manufacture of a membrane by phase inversion requires the dissolution of the constituent polymer of the future membrane and additives in a mixture of solvent(s): we obtain collodion. The triggering of the phase inversion is induced by a change in the thermodynamic state of the collodion. This change can be done in several ways: by temperature change TIPS (Temperature Induced Phase Separation) or by Rajabzadeh et al. [3] change of concentration in contact with a non-solvent of the polymer VIPS (Vapor Induced Phase Inversion) or not NIPS (Solvent Induced Phase Separation). After various solvent/non-solvent exchanges. The polymer concentration in the collodion solution is one of the important factors affecting the structure of the MD membrane. As the polymer concentration increases, the pore size and porosity decrease. Thus, the MD permeate flux becomes smaller using membranes prepared with higher polymer concentrations. It should be emphasized that the manufacture of MD membranes by the phase inversion technique from polymer solutions with a very low polymer concentration is practically impossible. For example, when the polymer content is below a threshold value (e.g. 10 wt% for PVDF, the resulting membranes become incoherent and holey, which can be observed clearly in the presence of light) [4].

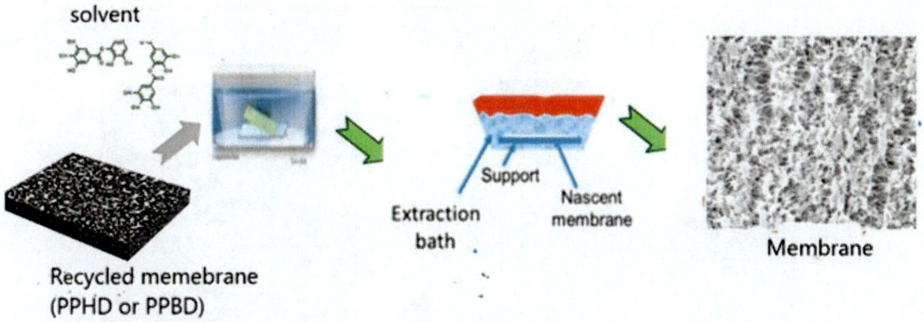

Fig. 1 Membrane preparation

2 Experimental Study

2.1 Preparation of Flat Membranes

In order to obtain flat membranes, the polymer solution called “collodion” is first prepared [4]. A mixture of butyl acetate and liquid paraffin was used as the solvent for the preparation of all the collodions; see Fig. 1. The mixture of polymers (PE, PP) is dissolved by adding the alumina powder with different quantities in the solvent under magnetic stirring in an oil bath at 130 °C until complete dissolution of the polymer. Once the collodion is prepared, it is spread on a glass plate using a manual film applicator. The resulting film is exposed to atmospheric conditions for 5 min then immersed in hexane then ethanol in order to extract the solvent from the membrane after it has been dried in an oven at 70 °C. Finally, the membrane is left for 24 h at ambient temperature before the characterization analyses.

The membrane surface was 8 cm^2, and the measurements were evaluated after a stabilization time of 40 min (time required to reach the state of equilibrium) at a transmembrane pressure of 1 bar and at room temperature (25 °C). After the stabilization period, the water permeate was collected within 60 s after the application of three different transmembrane pressures (the stabilization period from one to the other was approximately 10 min). Filtration experiments were performed with the same cross-flow filtration setup used for water permeability testing at 25 °C under pressures ranging from 0.6 to 1 bar depending on water permeability of the membrane (Fig. 2).

2.2 Determination of Membrane Morphology by SEM

Morphology was determined by scanning electron microscopy (SEM; Zeiss EVO, MA100, Assing, Italy) (Fig. 3). Cross-section specimens were fractured in liquid nitrogen, to prevent electrical charging of the membrane and to increase imaging

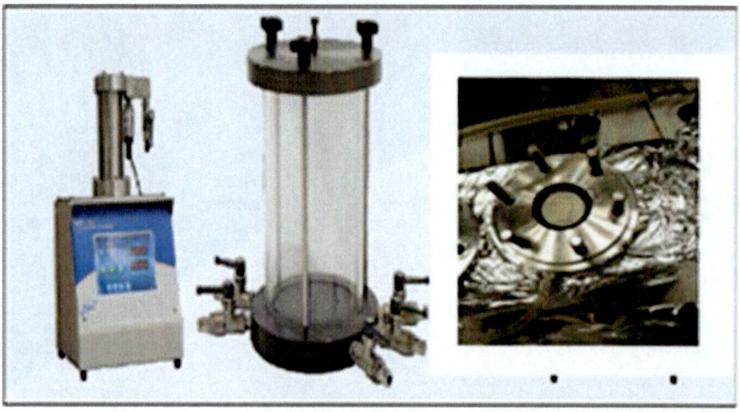

Fig. 2 Cross-flow cell used for permeability test

Fig. 3 Sample preparation for characterization with SEM

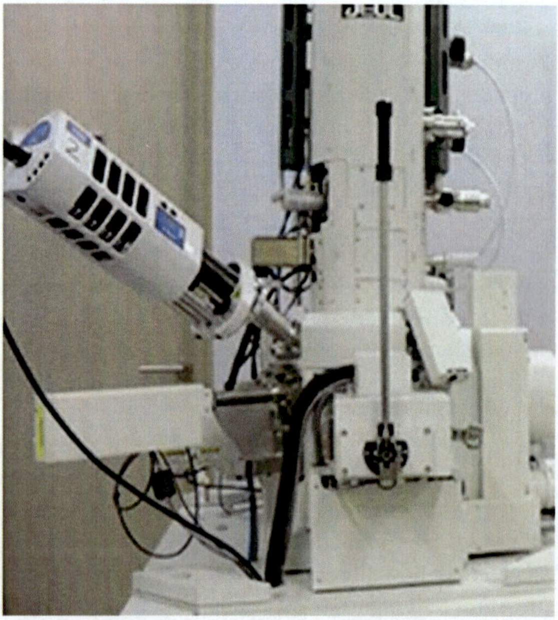

resolution, and mounted vertically on a [5] specimen holder. All samples (top, bottom, cross sections) were gold coated by a sputtering machine (Quorum Q 150R S).

Fig. 4 Micromètre numérique

2.3 *Thickness, Porosity, and Contact Angle*

Membrane thickness was detected by a digital micrometer (Carl Mahr, Göttingen, Germany) with an accuracy of ±0.001 mm and recorded in eight regions of each membrane. The membrane porosity values were calculated for 24 h, using the gravimetric method and considering the weight of the membrane samples before and after immersion in Isopropyl. The wettability of the membranes was analyzed by means of a contact angle instrument (CAM100 Instrument, Nordtest srl, GI, Serravalle Scrivia (AL), Italy) via the drop method, using droplets of ultrapure water (5 μL) placed on the membrane with a micropipette. For each membrane [6], five measurements were acquired and means and standard deviations were calculated. We can therefore deduce the hydrophilic or hydrophobic character of a surface (Fig. 4).

2.4 *FTIR*

In this part we are interested in studying the feasibility of preparing a porous hydrophobic membrane from recycled polymers. By examining the infrared spectra of pure and recycled HDPE and PP, it is clear that the spectra are almost identical. It can be seen that the variations of the peaks of the functional groups of the polymer before, after the recycling, and also by adding alumina are practically the same.

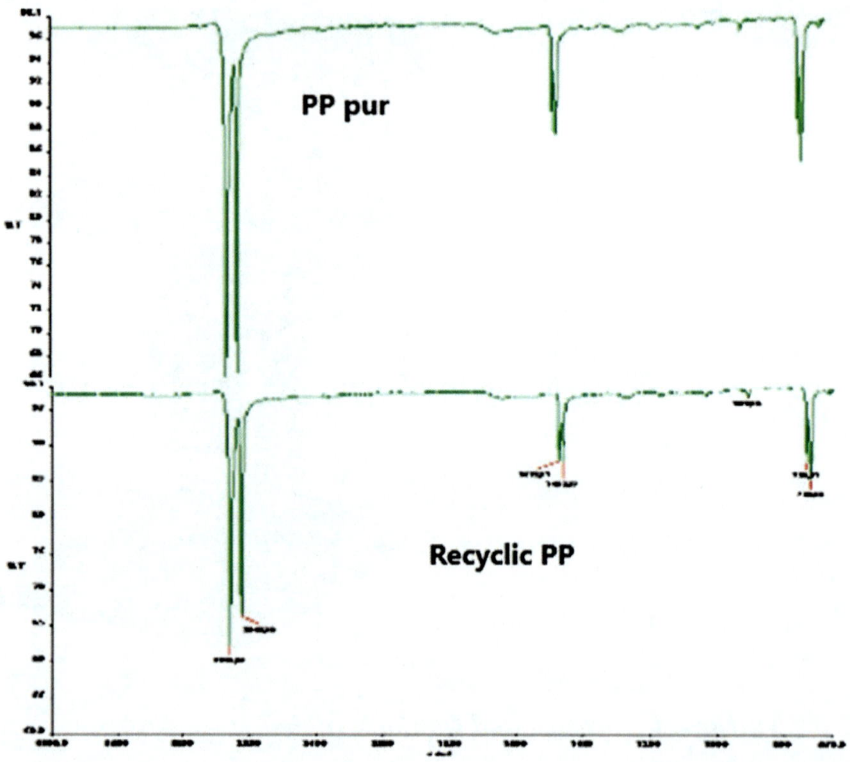

Fig. 5 FTIR analysis

Therefore, the polymer retains its properties after recycling and also after their use in the manufacture of membranes. Indicative spectra of prepared HDPE and spacer membranes are shown in Fig. 4. And the spectrum of spacer (PP) before reuse in the preparation of the membranes is represented in Fig. 5.

2.5 *Stability Test*

Several solubilization experiments at different temperatures were carried out using butyl acetate solution as solvent in the preparation of membranes from PE and PP. It was possible to obtain a homogeneous solution for the two polymers, but at a high temperature of 120 °C (Fig. 6).

Fabricated Membrane Validity Test: the membrane used in this unit is a hydrophobic polypropylene flat sheet membrane. These dimensions are 11 cm × 50 cm with an area of 0.055 m^2 (Fig. 7).

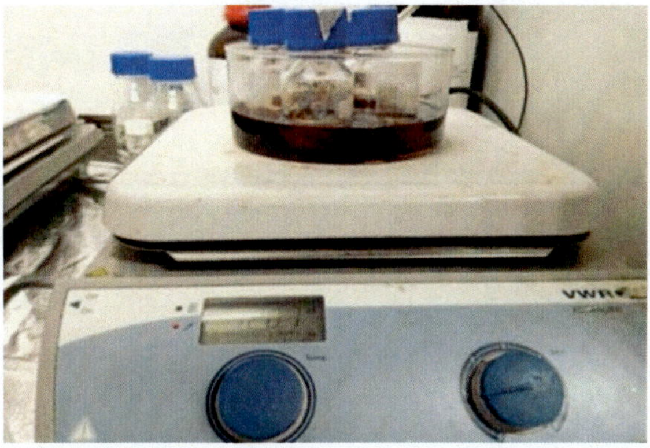

Fig. 6 Solubility test

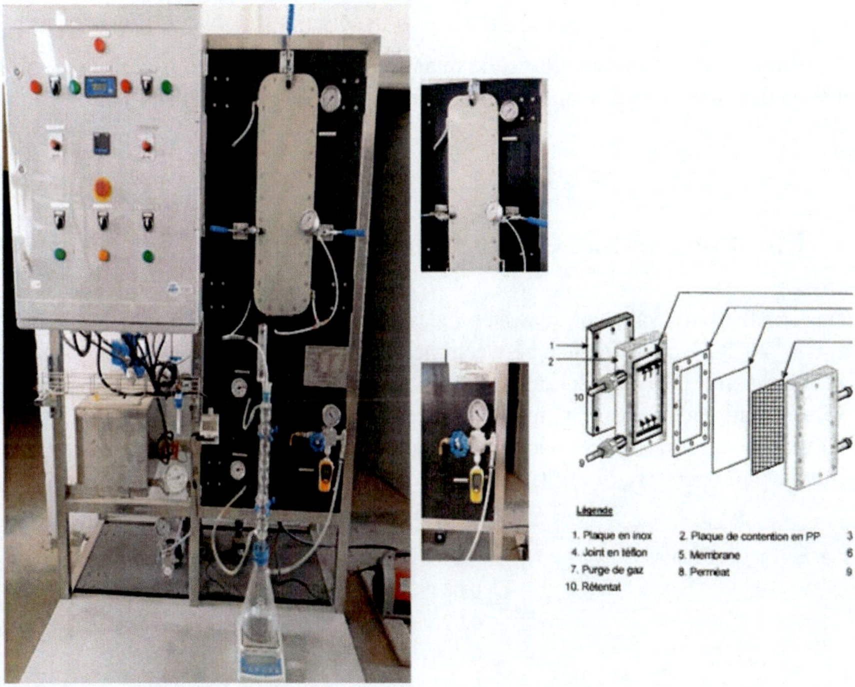

Fig. 7 Pilote setup

Table 1 Variation of sea water temperature

Salinity of sea water (g/L)	8	8	8	8	8
Sear water temperature (°C)	**50**	**60**	**70**	**80**	**90**
Flow water sea (L/h)	95	95	95	95	95
Pressure inlet of membrane (mbar)	400	400	400	400	400
Permeat pressure (mbar)	350	910	910	930	950
Cooling inlet of temperature (°C)	10	10	10	10	10
Cooling outlet of temperature (°C)	15	15	15	15	15
Flow of cooling water (L/h)	200	180	180	160	140
Flow of condensate water (L/h m^2)	**1.5**	**3.17**	**6.88**	**10.22**	**18.24**

3 Sea Water Temperature

The sea water temperature at the supply inlet is taken in the range [50–90 °C]. For the different study temperatures, the quantity of water obtained is measured. Note that an increase in inlet temperature results in an increase in condensate flow (Table 1).

4 Effect of Thermal Conductivity

The conductivity value of seawater measured in the laboratory is 0.04–0.06 W/m K at 20 °C. The difference between the value measured during on-site sampling and this value is solely due to the equipment and the measurement temperature. Subsequently, all measurements will be performed with the same device. Figure 8 shows the variations in conductivity according to the different water salinities at a reference temperature of 20 °C.

Fig. 8 Thermal conductivity

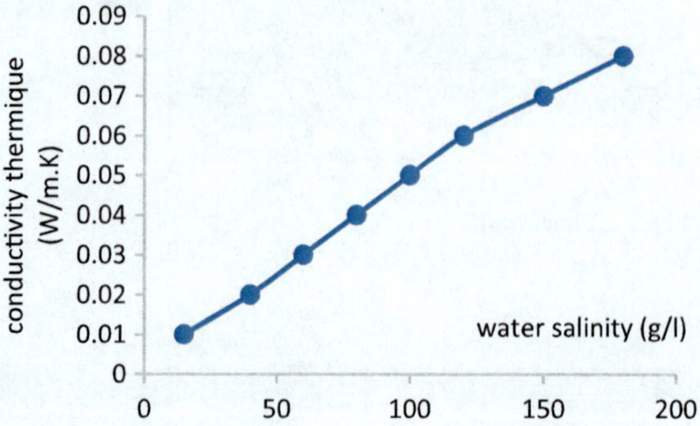

5 Conclusion

Membrane distillation is a promising technology for separation and purification processes. It is a thermally driven separation process in which only the vapor molecules pass through a porous and hydrophobic membrane. The results obtained showed that by increasing the feed temperatures, the flow of permeate increases. Regarding the effect of pressure, it appears that by increasing the latter, the permeate flux decreases while increasing salinity and thermal conductivity, the permeate flux increases. The thermal conductivity increases by increasing the salinity of the water.

References

1. Figoli A, Marino T, Simone S, Di Nicolò E, Li XM, He T, Tornaghi S, Drioli E (2014) Towards non-toxic solvents for membrane preparation: a review. Green Chem 9:4034–4059
2. Wu L, Sun J (2015) An improved process for polyvinylidene fluoride membrane preparation by using a water soluble diluent via thermally induced phase separation technique. Mater Des 86:204–214
3. Rajabzadeh S, Maruyama T, Sotani T, Matsuyama H (2008) Preparation of PVDF hollow fiber membrane from a ternary polymer/solvent/nonsolvent system via thermally induced phase separation (TIPS) method. Sep Purif Technol 2:415–423
4. Asma H, Khaoula H, Bechir C (2022) Comparative study of the performance of a locally manufactured membrane and the commercial one in vacuum distillation (VMD) of brackish water. https://doi.org/10.5004/dwt.2022.28029
5. Rasool MA, Pescarmona PP, Vankelecom IFJ (2019) Applicability of organic carbonates as green solvents for membrane preparation. ACS Sustain Chem Eng 16:13774–13785
6. Asma H, Khaoula H, Bechir C (2021) Manufacture of hydrophobic membranes using recycled polymers for the brackish water distillation, 11th International Renewable Energy Congress IREC. https://doi.org/10.1109/IREC48820.2020.9310400

Study of Solar Heating Process of Greenhouse in the Gabès Region

N. Anayed, A. Benhmidene, B. H. Assadi, K. Hidouri, B. Chaouachi, R. Boukhchina, and M. S. Belkadhi

Abstract The Tunisia has a great potential of solar thermal energy. This clean energy can be a good alternative to geothermal energy. In this context, our work is integrated, which aims to practice the solar thermal energy for the continuous heating of an agricultural greenhouse. To achieve this objective, we have installed a heating system consisting of 40 thermal solar panels which are used to heat the water that will be stored in a 15 m^3 tank during the day. At night, the stored hot water is used to heat an agricultural greenhouse with an area of 300 m^2. The indicated setup was placed in the experimental site of the Technical Center of Protected and Geothermal Crops located in the Gabes region at longitude 10.097 and altitude 33.888. The climate parameters of indoor air such as relative humidity and temperature were flowing against time and compared the ambient air and it is of the unheated greenhouse. A campaign to measure the different climatic parameters for 8 successive days (25/02/2020 until 03/03/2020) has been started. Les résultats expérimentaux du suivi montrent qu'une augmentation de 5 à 6 °C la température de l'air de la serre chauffée par rapport à la serre non chauffée. Le même comportement est obsервé pour l'humidité relative qui croit de 70 à 80%.

Keywords Solar energy · Renewable energy · Indoor temperature · Relative humidity · Greenhouse

---

N. Anayed (✉) · A. Benhmidene · K. Hidouri · B. Chaouachi
Laboratory of Energy, Water, Environment and Processes, The National School of Engineering of Gabès, Gabès University, Gabes, Tunisia
e-mail: anayed.naima@gmail.com

R. Boukhchina
Laboratory of Eremology and Combating Desertification (LR16IRA01), Arid Regions Institute, University of Gabes, 4119 Medenine, Tunisia

B. H. Assadi · M. S. Belkadhi
Dryland and Oases Cropping Laboratory, Arid Regions Institute, Street El Jorf, 4119 Medenine, Tunisia

N. Anayed · B. H. Assadi · M. S. Belkadhi
Technical Center for Protected and Geothermal Crops, Gabes, Tunisia

R. Khiari and M. Jawaid (eds.), *Proceedings of the 3rd International Congress of Applied Chemistry & Environment (ICACE-3)*, Springer Proceedings in Materials 23, https://doi.org/10.1007/978-981-99-1968-0_16

1 Introduction

In southern Tunisia, greenhouse cultivation presents main sector that have experienced significant development. This progression is mainly technological and economic since greenhouse crops are mainly intended for export. This development is the result of the significant increase in demand for early crops [1]. In Tunisia, the export period for greenhouse products coincides with the winter period. To ensure a product that meets export requirements, microclimate management is necessary [2].

But these different types of greenhouses found in the Mediterranean basin in an arid to semi-arid climate promotes very painful climatic parameters for the development of greenhouse crops especially in the winter period is very commonly reflected in the inversion of temperatures between the 'Interior and exterior and by great heterogeneity between daytime and night temperature and other climatic complications which mitigates the development of solanaceae and cucurbit, reduces the yield and profitability of crops and promotes the development of cryptogamic cultures which put obviously the importance of managing climatic parameters inside the greenhouse [3].

A good comprehension, modelling, and simulation of climatic parameters with all these elements (temperature inside and surrounding, relating humidity, exchanges radiatifs, and convectifs) is primordial to know how to make sure and to maitriser these parameters, while pointing out that every climatic parameter plays a significant role [4]. In another way, the sudden and successive increase in the cost of energy have led several researchers in the world, where there are large areas of greenhouses, to think about cheaper heating alternatives and to use microclimate measuring devices to properly evaluate the different parameters.

In Tunisia and mainly in southern Tunisia which is known for its arid to semi-arid climate, early crops are experiencing a remarkable precocity compared to greenhouse growers in the world given the importance of solar radiation in these areas and the physicochemical quality of geothermal waters which is the main source for irrigation which makes it interesting to study the energy behavior of the microclimate in greenhouses using a simple and easy to use a simple and easy to use model [5]. Treat to estimate the energy needs of an agricultural greenhouse [6].

In this context, our study is part of heating a multichapel greenhouse type Filclair with an area of 300 m^2 sited at the experimental station of the Technical Center for Protected and Geothermal Crops using 40 thermal solar panels and an isothermal basin for the storage of hot water.

2 Materials and Methods

2.1 Experimental Setup

The solar thermal unit that is used to heat an agricultural greenhouse is essentially composed of:

- Agricultural greenhouse
- Heating system
- Isothermal storage tank.

2.1.1 Agricultural Greenhouse

It is a chapel greenhouse type Filclair of area 300 m^2 and height under gutters 3.5 m intended mainly for the planting of nightshades and cucurbits. This greenhouse is covered mainly by a thermal plastic film with diameters of 200 μm of transparent color.

Studied greenhouse is equipped with an internal heating network consisting of 11 loops of length 65 m laid in double loops near the root system.

The technical characteristics of this greenhouse are given in Table 1.

2.1.2 Heating System

It consists of 40 thermal solar panels connected in batteries of 5 in series, so in total we have 8 batteries in series. Hot and cold water circulates through copper pipes of different diameters via circulation pumps (Fig. 1).

2.1.3 Isothermal Tank

It is a concrete basin, and these dimensions are defined in Fig. 2; it is insulated by polystyrene. The technical characteristics of this basin are given in the following table.

Table 1 Characteristics of the greenhouse

Greenhouse elements	Values
Cultivation's covered area	2 × 300 m^2
Number of aisles	02
Width of aisle	9.6 m
Total width	19.2 m
Length	31 m
Height under crop support	3.5 m

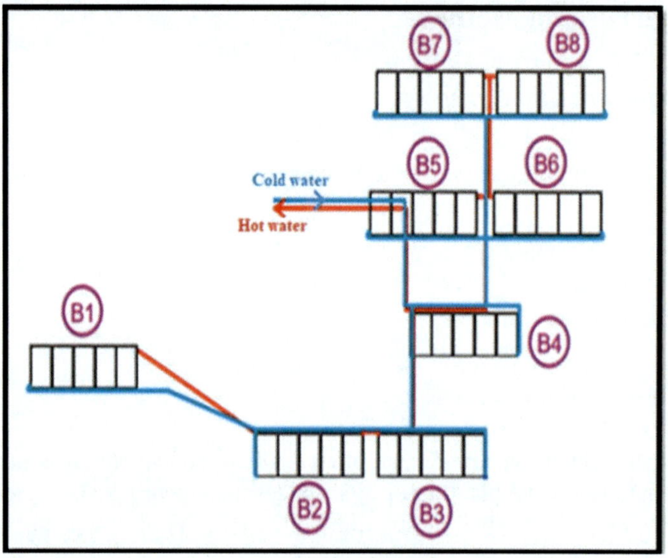

Fig. 1 Location of solar panels

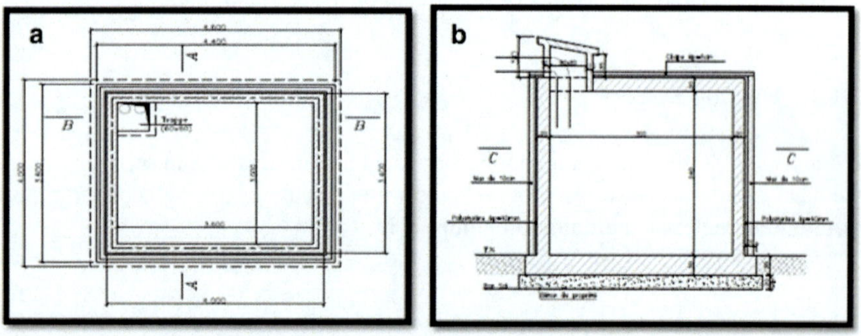

Fig. 2 **a** Cut formwork plan C–C, **b** cut formwork plan A–A

2.2 *Measuring Equipment*

The equipment used to monitor the climatic parameters inside the pilot greenhouse consists essentially of two humidity and temperature recorders of type HMI 172 SI. They are used to give successive measurements every 30 min.

Outdoor climate measurements are given by a weather station type IMETOS [7] (Fig. 3).

Fig. 3 Main components of experimental set-up: **a** greenhouse, **b** weather station, **c** thermal panels, **d** pumps circulation water, **e** tank, and **f** heating system

2.3 *Measuring Principle*

The hot water circulates in a system shown in Fig. 4. This circulation is done via a Salmon type pump. It is done between the thermal panels and the storage tank, during the day. However; the circulation will be between the tank and the greenhouse during the night. This is done by using a second pump with the same characteristics as the first one (Fig. 5).

The measurement principle for this campaign to monitor climatic parameters which began on February 25, 2020 until March 03, 2020 is defined as follows: a humidity and temperature recorder was installed inside the pilot greenhouse precisely in the middle and 2 m high near the apex of the tomato crop [8].

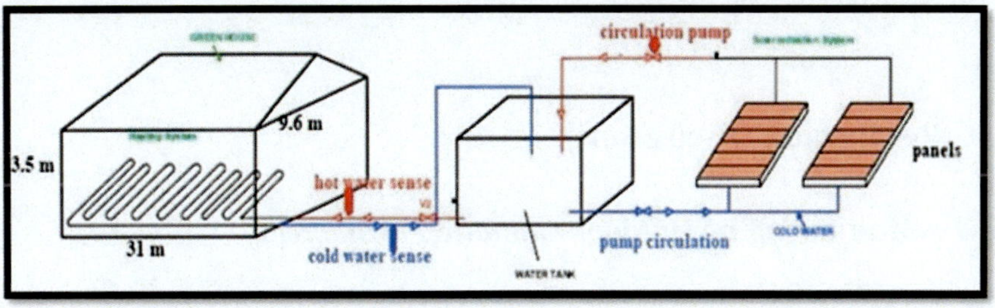

Fig. 4 Principle of operation of the solar unit [8]

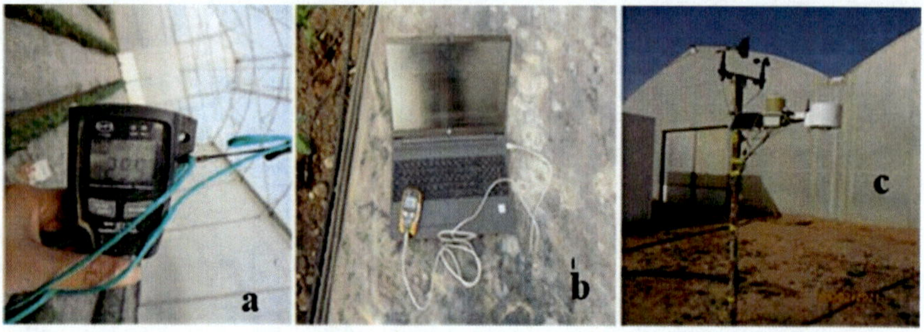

Fig. 5 **a** Humidity and temperature logger, **b** data recording, **c** weather station brand IMETOS

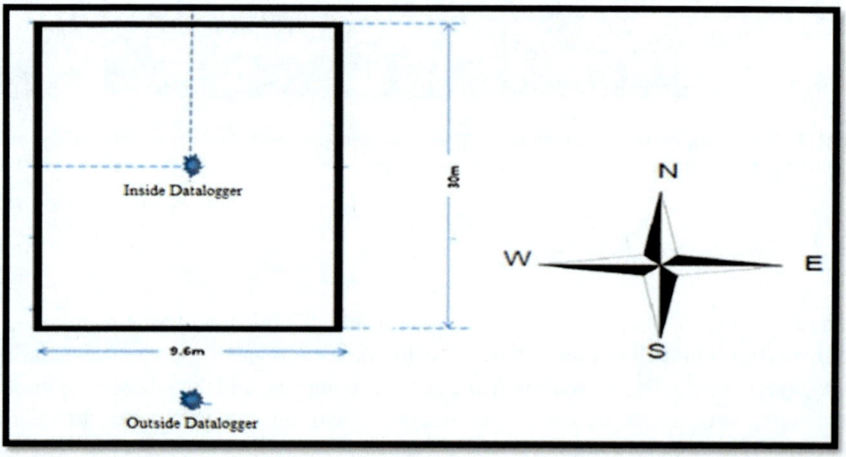

Fig. 6 Location of measuring devices

The other recorder is installed outside almost 3 m in front of the greenhouse to record the outdoor climatic parameters, in total, 48 relative humidity and temperature values are obtained by each recorder (Fig. 6).

3 Results and Discussions

3.1 Monitoring of Outdoor Climatic Parameters

We started this measurement campaign by monitoring the external climatic parameters (wind speed, relative humidity, precipitation, radiation, and ambient temperature), and we obtained the figures below.

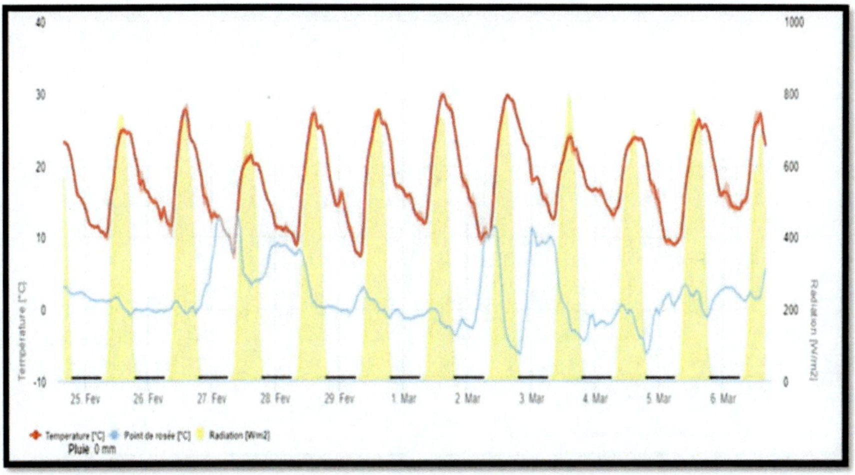

Fig. 7 Variation of radiation and ambient temperature during the measurement campaign

From Fig. 7 above, it can be seen that the daily radiation during the measurement campaign is in the vicinity of 780 W/m^2 which is a very favorable measure to improve the microclimate under greenhouse using solar thermal heating.

The night ambient temperature is below 10 °C for several overnight stays, which makes it important to heat the greenhouse overnight to ensure the ideal temperature for the progress of the tomato crop and eliminate the various pests.

Figure 8 shows the variation of the wind speed during the measurement. It's easy to see that the average speed of the prevailing wind is between 0.5 and 1.5 m/s along the day and very small at night, so important precautions must be taken to protect yourself from this harmful wind for the greenhouse microclimate.

In the first place it is necessary to eliminate as much as possible the opening of the openings of the multichapel greenhouse (the aeration must be very low) so as not to lower the relative humidity and in the second place it is also necessary to heat during the beginning of the day and to control the RH frequently so as not to develop fungal diseases sensitive to decreases in humidity.

3.2 *Monitoring of the Tank and Panel Temperatures*

In the aim to decrease the heat losses, the heating network linking the thermal collectors and the storage tank was thermally insulated. A series of measurements of humidity and the temperature inside the greenhouse was started on February 25, 2020 and was completed on March 03, 2020.

Figure 9 shows that the output maximum panel temperature can reach 69 °C while the maximum storage tank temperature is around 55 °C. Where, it's obtained around

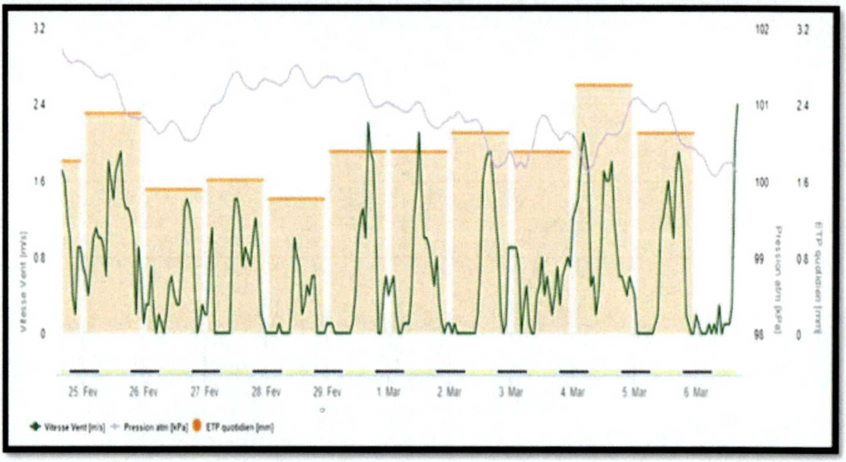

Fig. 8 Variation of wind speed and evapotranspiration during the measurement campaign

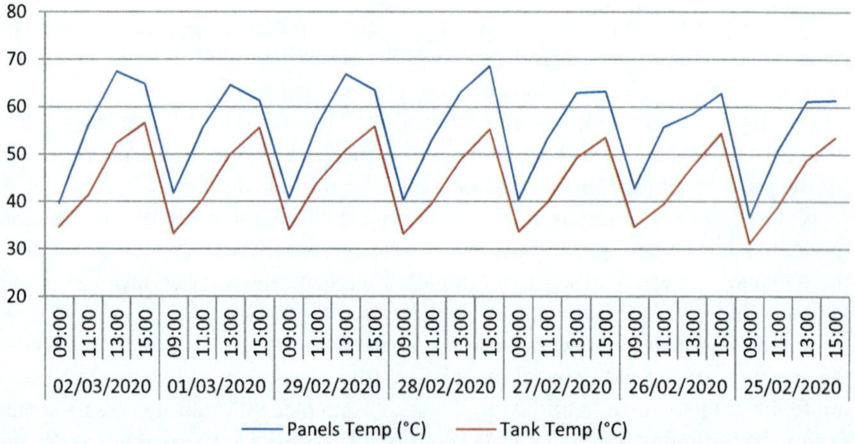

Fig. 9 Variation of panels and tank temperature during the measurement campaign

3 p.m. It decreases by 2–3 °C around 7 p.m., when the heating process starts. The temperature of the heating water reaches a temperature between 35 and 30 °C after one night of heating. This result shows the efficiency of the adopted heating system.

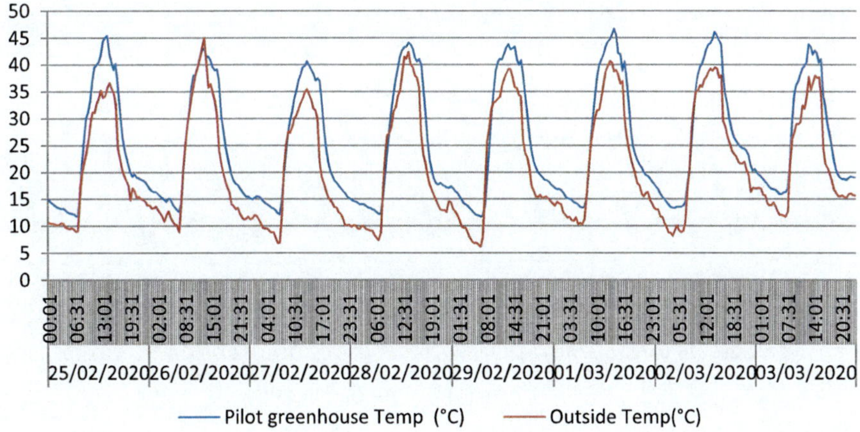

Fig. 10 Variation of the greenhouse temperature according to the outside temperature

3.3 *Improvement of the Air Temperature Inside Greenhouse*

The temperature inside the greenhouse compared to the outside is shown in Fig. 10. We note that the heat gain inside the greenhouse is around 5 °C and can reach 7 °C for a sunny day.

These recorded values show the importance of managing a microclimate in a multispan greenhouse to improve the yield of different crops planted.

During this measurement campaign, it is noticed that the temperature inside the heated greenhouse is higher than the ambient temperature, especially during the night period [9]

3.4 *Improvement of the Relative Humidity Inside Greenhouse*

For the same measurement campaign, a recording of the relative humidity inside the heated greenhouse and outside is recorded in Fig. 11.

A relative humidity of around 80% is recorded for the majority of the measurement days with a very low diurnal relative humidity also around 20%, which requires the intervention of other techniques to humidify (misting, aeration…).

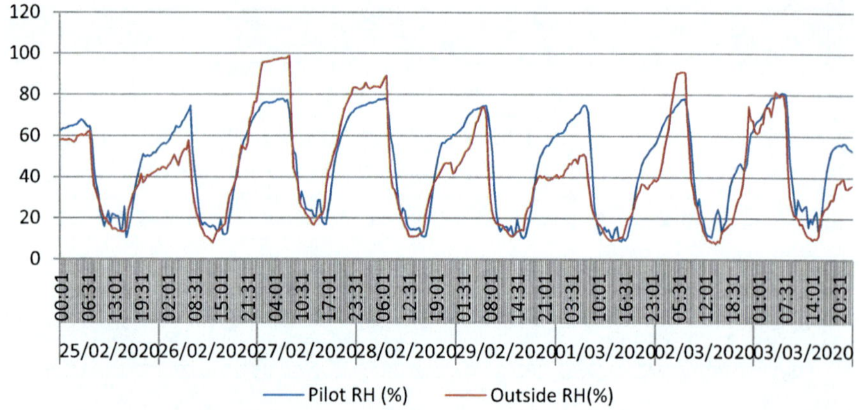

Fig. 11 Variation of the relative humidity of the pilot greenhouse according to the outside RH

4 Conclusions

The present experimental study focuses on the valorization of solar thermal energy for heating agricultural greenhouses. An agricultural greenhouse with an area of 300 m^2 located at the Chenchou region, Gabes in Tunisia is used.

The heating system is mainly composed of thirty thermal solar panels to heat the water, an isothermal basin with a volume of 15 m^3 used for the storage of hot water during the day. And during the night, hot water is used to heated greenhouse already described via an internal heating network composed mainly of ringed polypropylene loops. The temperature of the water at the outlet of the thermal panels reaches 65 °C. In the other hand, the maximum value of temperature water inside storage tank is 55; this allows the air temperature inside greenhouse to reach 14 °C. A gain of 7 °C is obtained compared to no heated greenhouse and 9 °C compared of the outside.

In the same way, the relative humidity of heated greenhouse is stabilized between 60 and 80%. This represents the most favorite interval for the development of crops. In the absent of the heating process relative humidity reached 80%.

References

1. Bazgaou A, Fatnassi H, Bouharroud R, Ezzaeri K, Gourdo L, Wifaya A, et al (2021) Effect of active solar heating system on microclimate, development, yield and fruit quality in greenhouse tomato production. Renew Energy 165:237–250. https://doi.org/10.1016/j.renene.2020.11.007
2. Ezzaeri K, Fatnassi H, Wifaya A (2020) Performance of photovoltaic Canarian greenhouse: a comparison study between summer and winter seasons. Sol Energy 198:275–282. https://doi.org/10.1016/j.solener.2020.01.057
3. Gourdo L, Fatnassi H, Bouharroud R (2019) Heating Canarian greenhouse with a passive solar water–sleeve system: effect on microclimate and tomato crop yield. Sol Energy 188:1349–1359. https://doi.org/10.1016/j.solener.2019.07.004

4. Barbaresi A, Maioli V, Bovo M (2020) Application of basket geothermal heat exchangers for sustainable greenhouse cultivation. Renew Sustain Energy Rev 129:109928. https://doi.org/10.1016/j.rser.2020.109928
5. Besser H, Younes H (2019) Causes and risk evaluation of oil and brine contamination in the Lower Cretaceous Continental Intercalaire aquifer in the Kebili region of southern Tunisia using chemical fingerprinting techniques. Environ Pollut 253:412–423. https://doi.org/10.1016/j.envpol.2019.07.020
6. Maaoui M, Boukchin R, Hajjaji N (2021) Environmental life cycle assessment of Mediterranean tomato: case study of a Tunisian soilless geothermal multi-tunnel greenhouse. Environ Dev Sustain 23(2):1242–1263. https://doi.org/10.1007/s10668-020-00618-z
7. Assadi BH, Chouikhi S, Ettaib R (2021) Effect of the native strain of the predator Nesidiocoris tenuis Reuter and the entomopathogenic fungi Beauveria bassiana and Lecanicillium muscarium against Bemisia tabaci (Genn.) under greenhouse conditions in Tunisia. Egypt J Biol Pest Control 31(1):1–11
8. Anayed N, Benhmidène A, Chaouachi B (2022) The efficiency of a continuous solar system for heating a greenhouse in southern Tunisia. Int J Ambient Energy 1–9. https://doi.org/10.1080/01430750.2022.2031290
9. Neira DP, Montiel MS, Cabeza MD, Reigada A (2018) Energy use and carbon footprint of the tomato production in heated multi-tunnel greenhouses in Almeria within an exporting agri-food system context. Sci Total Environ 628:1627–1636. https://doi.org/10.1016/j.scitotenv.2018

Eco-Friendly Application of Mauritanian Limestone Clay: Malachite Green Removal from Textiles Wastewater Case Study

Youssef Aoulad El Hadj Ali, Abdoulaye Demba N'diaye, Kaouthar Benahdach, Amal Bouchta, Mohamed Sid'Ahmed Kankou, and Mostafa Stitou

Abstract The present work has been performed to evaluate the employability of natural limestone clay in the removal of Malachite Green (MG) from aqueous solution. Some parameters such as adsorbent dosage, pH, contact time, temperature, and initial concentration on the removal rate were studied. Adsorption kinetic data were analyzed using Pseudo First Order (PFO) and Pseudo Second Order (PSO) models. For the kinetic study, the adsorption process was fitted to the PSO model. The Langmuir, Freundlich, and Sips isotherms were applied to fitting the experiment data. The results indicate that the parameters of the isotherm models are appropriate to fit the current adsorption isotherm data in the following order: Langmuir > Sips > Freundlich. The maximum uptake amount q_m, was found to be 55.84 mg g^{-1}. The present work shows that the clay is a promising adsorbent for the removal of MG from aqueous solutions.

Keywords Cationic dye · Malachite green · Limestone clay · Adsorbent · Aqueous solution

Y. Aoulad El Hadj Ali · K. Benahdach · A. Bouchta · M. Stitou
Laboratoire de L'Eau, les Etudes et les Analyses Environnementales, Département de Chimie, Faculté des Sciences, Université Abdelmalek Essadi, BP 2121, Mhannech II, 93002 Tétouan, Maroc

A. Demba N'diaye (✉) · M. S. Kankou
Unité de Recherche Eau, Pollution et Environnement, Département de Chimie, Faculté des Sciences et Technique, Université de Nouakchott Al Aasriya, BP 880, Nouakchott, Mauritanie
e-mail: abdouldemba@yahoo.fr

R. Khiari and M. Jawaid (eds.), *Proceedings of the 3rd International Congress of Applied Chemistry & Environment (ICACE-3)*, Springer Proceedings in Materials 23, https://doi.org/10.1007/978-981-99-1968-0_17

1 Introduction

The environmental activists are concerned about the future and are sounding the alarm about the environmental pollution caused by the discharge of industrial effluents into the environment without adequate prior treatment [1]. The textile industry is classified among the first pollutants of the environment that used dyes to color their products and also consume large volumes of water [14].

However, the presence of dyes in effluents is a major concern because of their adverse effects on many life forms [20]. Malachite green (MG), a cationic dye, is widely used in industrial and pharmaceutical ways. Moreover, the MG is conventionally used as a fungicide in aquaculture [15]. As with major types of dyes, MG is an extremely harmful, mutagenic, teratogenic, and carcinogenic dye [9].

For all these reasons removing MG from industrial wastewater before discharging it becomes a necessity. Some processes such as electrochemical degradation [3], photo-degradation [17], and photocatalytic degradation [8] are used to remove MG from aqueous solution.

However, these techniques are very expensive and cannot treat a wide variety of wastewater containing dyes. Adsorption is a technique recognized as the most effective compared to other sophisticated methods, due to its ease of design, high performance, and cost-effectiveness [10]. In addition, the adsorbent-grade activated carbon is cost-prohibitive [13]. For this reason, the aim of this study is to investigate the possibility of using local clay for the removal of MG from aqueous solutions.

2 Materials and Methods

2.1 *Adsorbate and Adsorbent*

The stock solution of MG was prepared by dissolving 1 g of MG powder in 1 L of distilled water. All working solutions of desired initial MG concentrations were prepared by diluting the stock solution with distilled water. Other concentrations are prepared by dilutions and used to develop the standard curves using the Spectrophotometer Hach DR 2800.

2.2 *Preparation and Characterization of Clay Adsorbent*

The raw clay was collected from Hassi El Ebyed in the Moughataa of M'Bout, Wilaya of Gorgol, Mauritania. The sample was watched several times with demineralized water, then oven-dried at 105 °C until a stable weight was obtained, then ground in a mortar. The particles were separated according to their diameter and stored in a desiccator for later use. In addition, the limestone clay was characterized using several analysis methods such as FTIR, Scanning Electron Microscopy (SEM), X-ray Diffraction, X-ray Fluorescence, and pH_{PZC}.

2.3 Batch Adsorption Studies

MG adsorption using natural clay as an adsorbent process was conducted in batch experiments. The effects of system conditions, mass effect (0.02–1 g), pH effect (4, 7, and 10), contact time (0–120 min), and effect of temperature and initial concentration on the MG removal were evaluated for natural limestone clay. The adsorption isotherms were obtained by varying the initial MG concentrations from 0 to 60 mg L^{-1}. At the end of each experiment, the stirred solution mixture was centrifuged and the residual concentration of MG was analyzed by Spectrophotometer at 618 nm wavelength. The adsorption uptake at equilibrium time q_e (mg g^{-1}) and percentage of the MG removed (%) are expressed by following Eqs. (1) and (2), respectively:

$$q_e = \frac{(C_i - C_e)V}{m} \tag{1}$$

$$\text{Removal } \% = \frac{C_i - C_e}{C_i} \times 100 \tag{2}$$

where:

q_e is the MG concentration in adsorbent (mg g^{-1}),

C_i is the initial MG concentration (mg L^{-1}),

C_e is the MG concentration at equilibrium (mg L^{-1}),

V is the solution volume (L),

m is the mass of the natural clay as adsorbent used (g).

All batch experiments were conducted in triplicate and the average values are reported.

3 Results and Discussion

3.1 Characterization of Natural Clay Adsorbent

The elemental chemical analysis of the clay material used in this work was carried out by X-ray Fluorescence, the results of which are listed in Table 1.

Table 1 Chemical composition of the clay sample

Components	(%)
SiO_2	2.07
Al_2O_3	0.20
Fe_2O_3	0.16
CaO	53.88
K_2O	0.10
P_2O_5	0.62
MnO	0.002
TiO_2	0.03
SO_3	0.05
SrO	0.03
Cr_2O_3	0.001
CuO	0.006

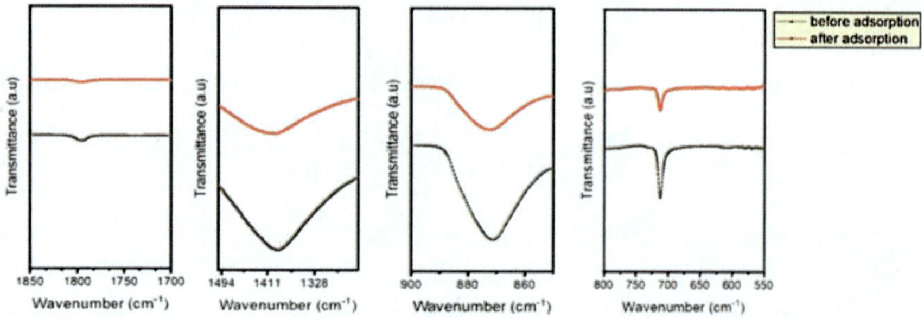

Fig. 1 FTIR spectra of natural clay adsorbent before and after MG adsorption

The results indicate that the material has a significant amount of calcite (53.88%) and Silica (2.07%). The overall composition of other oxides (Al_2O_3, Fe_2O_3, K_2O, P_2O_5, MnO, TiO_2, SO_3, SrO, Cr_2O_3 and CuO) considered as impurities reaches 1.2%.

Figure 1 shows FTIR spectra of the natural clay before and after MG adsorption. The result of the FTIR study of clay before adsorption showed absorption peaks located at 1801, 1405, 872, and 711 cm^{-1}. The small peak at 1801 cm^{-1} can be attributed to the presence of C=O elongation. The band at 1405 cm^{-1} related to C=O stretching of ketones, aldehydes, lactones or carboxyl groups. The bands located at 700 and 810 cm^{-1} are related to the deformation vibration of Al–Al–OH groups of smectites [12]. After MG adsorption, the slight shift of peaks was at 1805, 1423, 874, and 713 cm^{-1} (Fig. 1). A slight shift of frequency and a decrease in intensity of some peaks could be attributed to MG ions adsorbed to natural clay adsorbent. The FTIR spectroscopic study after adsorption confirmed that the MG has a binding ability with the surface of natural clay adsorbent.

The identification of the crystalline structures of limestone clay was carried out using a Bruker diffractometer (Phillips Xpert-pro model) with Cu Kα $\lambda = 1.54051$ Å radiation with a step of 0.017°/1 s. In addition, a semi-quantitative analysis of the mineralogical composition was performed using the Powder Cell 2.4 software D8 Advance 500 model.

Figure 2 presents the diffractogram of natural limestone clay; the peaks show the presence of quartz (SiO_2) and calcite ($CaCO_3$) in the majority and the presence of (Fe_2O_3) as a trace. SEM image of the limestone clay powder was performed using a "Hirox scanning electron microscopy". Analysis of SEM micrographs (Fig. 3) of the calcareous clay samples showed smoother and more uniform surface characteristics with the existence of a few pores in the adsorbent surface.

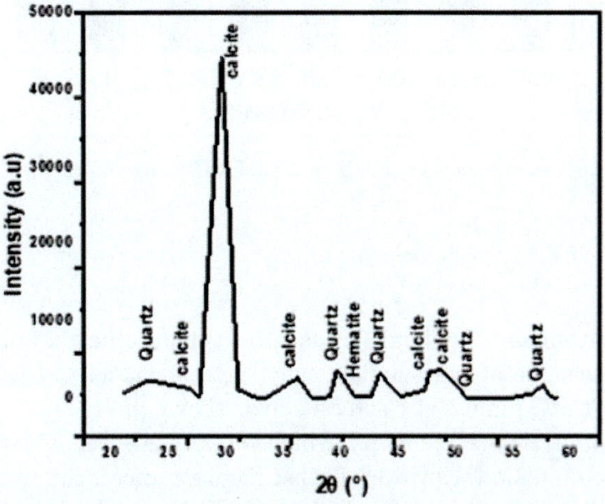

Fig. 2 DRX of limestone natural clay

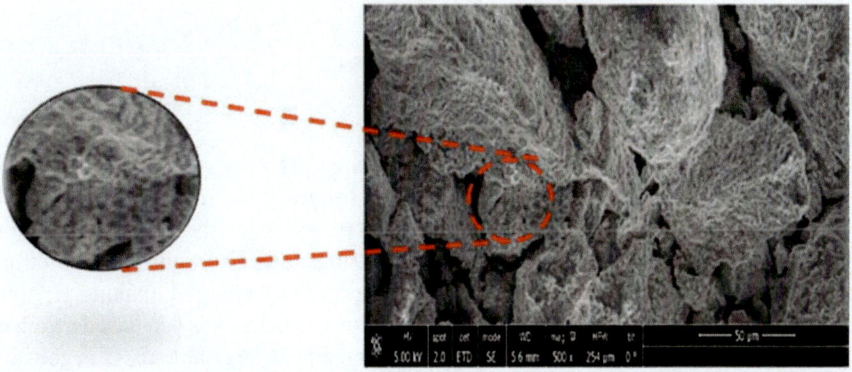

Fig. 3 SEM micrographs of limestone clay surface before adsorption

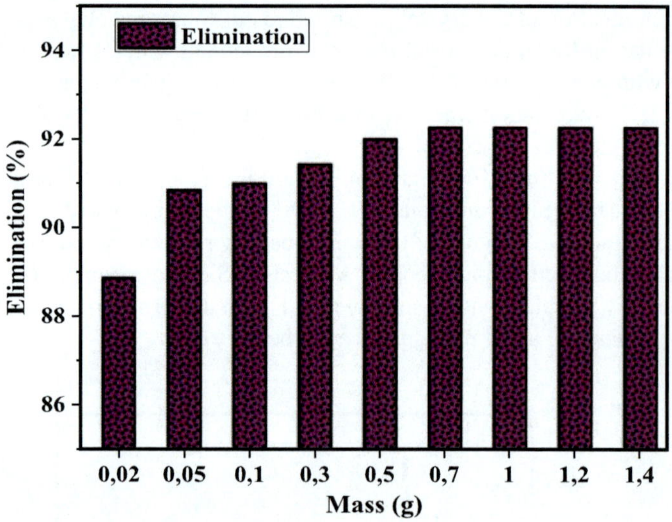

Fig. 4 Effect of adsorbent dosage on the adsorption of MG by natural clay

3.2 Effect of Clay Adsorbent Mass

The effect of the amount of natural clay adsorbent on the efficiency of adsorption was also studied. Variation of doses in the range 0.02–1.4 g at a fixed MG concentration (10 mg L^{-1}) for MG removal by natural clay is shown in Fig. 4.

The results indicate that the increase in the dose of adsorbent leads to an increase in adsorption elimination. However, further increase after a certain adsorbent mass does not improve the adsorption which may be due to the interference between binding sites of the natural clay adsorbent at different doses. The optimal natural clay adsorbent dose obtained in this work is 0.7 g.

3.3 Effect of pH

The effect of pH in this work was studied at three pH values (4, 7, and 10) using 10 mg L^{-1} of MG solution. As seen in Fig. 5, the solution pH affects the MG removal, and therefore, the removal efficiency of MG increased when the pH of the solution decreased from 75.4% at pH 7 to 64.9% at pH 4. In acidic conditions of the solution, the surface charge of the natural clay became positively charged, making H^+ ions compete with cationic MG dye. Thereby the removal efficiency is decreased owing to electrostatic repulsion force. While at pH higher than 2, a large amount of negatively charged OH^- found on the surface of natural clay results in an enhanced electrostatic attraction, which facilitates greater MG adsorption capacity. However, the removal efficiency of MG increased when the pH of the solution increased from 75.4% at pH 7

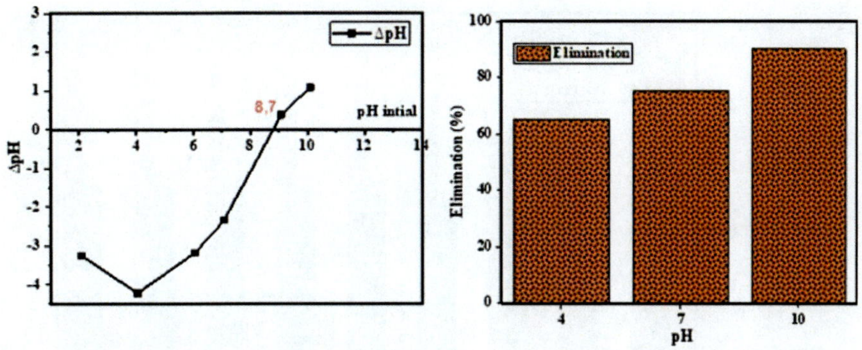

Fig. 5 Point of zero charge of limestone clay and the effect of pH on adsorption of MG onto limestone clay

to 90% at pH 10. The combined effect of both chemical and electrostatic interactions between the oxide surface and MG could be responsible for the adsorption reduction.

3.4 *Effect of Contact Time*

The effect of contact time and the initial concentration on the removal of MG (10–30 mg L^{-1}) onto clay are shown in Fig. 6. The results of the contact time effect have shown that the adsorption capacity increased with the increase of contact time and reach equilibrium in 60 min for all concentrations. The initial faster rates of adsorption may also be attributed to the presence of a large number of binding sites. Similar behavior was obtained for the adsorption of MG by neem sawdust [5].

3.5 *Study of Adsorption Kinetics*

Pseudo First Order (PFO) and Pseudo Second Order (PSO) models were applied to investigate the adsorption data. The linear kinetics PFO and PSO models expressed by (3) and (4), respectively:

$$\log(q_e - q_t) = \log q_e - \frac{k_1}{2.303}t \quad (3)$$

$$\frac{t}{q_t} = \frac{1}{k_2 q_e^2} + \frac{t}{q_e} \quad (4)$$

where:

q_t is the amount of MG adsorbed per unit mass of natural clay (mg g^{-1}) at time t,

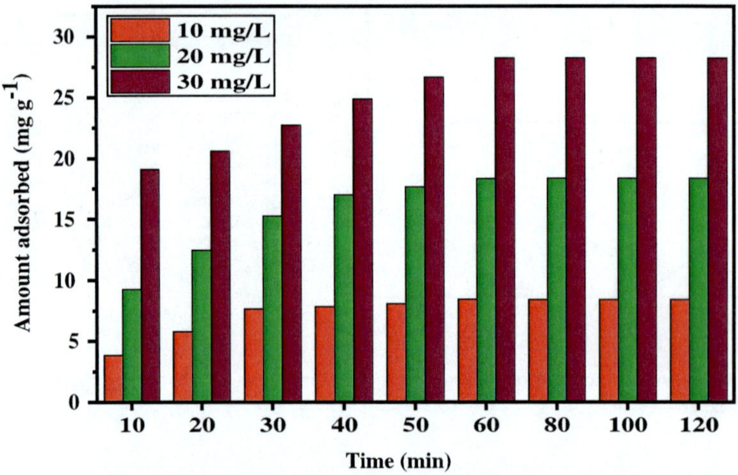

Fig. 6 Kinetics of MG adsorption by natural clay for various initial dye concentrations

k_1 (L min^{-1}) is the PFO rate constant,

k_2 (mg g^{-1} min^{-1}) is the PSO rate constant for adsorption,

q_e (mg g^{-1}) the amount of MG adsorbed at equilibrium,

t is the contact time (min).

The adsorption kinetic parameter values and R^2 values were calculated using the linear method by Excel. The calculated values were listed in Table 2. The fitting of the PFO model was found not to be satisfactory ($R^2 = 0.906$–0.992), while it was excellent ($R^2 = 0.994$–0.997) for the PSO kinetic model suggesting that chemisorption is the rate controlling step. Similar kinetics was observed in adsorption MG on Effective microorganisms-based compost [11].

Table 2 Linear kinetic model parameters

Model	Parameters	10 mg L^{-1}	20 mg L^{-1}	30 mg L^{-1}
	q_{exp}	8.44	18.39	28.23
PFO	q_e	6.89	15.44	18.57
	k_1	0.062	0.088	0.082
	R^2	0.906	0.992	0.950
PSO	q_e	9.32	20.32	30.39
	k_2	0.106	0.103	0.129
	R^2	0.994	0.996	0.997

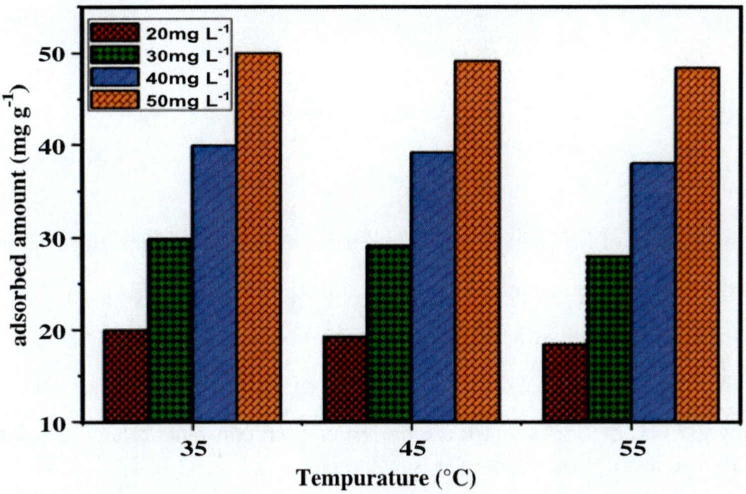

Fig. 7 Effect of temperature on MG adsorption by clay

3.6 Effect of Temperature and Initial Concentration on the MG Removal

Figure 7 shows the effect of temperature on MG adsorption by varying the temperature in the range of 35–55 °C with MG concentrations of 20, 30, 40, and 50 mg L^{-1}. It was seen that the adsorption decreased with increasing temperature for all the initial concentrations, suggesting that the adsorption of MG onto the adsorbent is an exothermic process.

3.7 Adsorption Isotherm

In this study, the Langmuir, Freundlich, and Sips models were used for describing the experimental results. The Langmuir adsorption isotherm assumes that adsorption takes place at specific homogeneous surface sites within the adsorbent [6]. The Freundlich isotherm is employed to describe the heterogeneity of the adsorbent surface [2]. The Sips isotherm is a combination of the Langmuir and Freundlich isotherms, which represent systems for which one adsorbed molecule could occupy more than one adsorption site [16]. The non-linearized isotherm Langmuir, Freundlich, and Sips models expressed by Eqs. (5), (6), and (7), respectively:

$$q_{\mathrm{e}} = \frac{q_{\mathrm{m}} K_{\mathrm{L}} C_{\mathrm{e}}}{1 + K_{\mathrm{L}} C_{\mathrm{e}}} \tag{5}$$

$$q_{\mathrm{e}} = K_{\mathrm{F}} C_{\mathrm{e}}^{1/n} \tag{6}$$

$$q_{\mathrm{e}} = q_{\mathrm{m}} \frac{K_{\mathrm{S}} C_{\mathrm{e}}^{n}}{\left(1 + K_{\mathrm{S}} C_{\mathrm{e}}^{n}\right)} \tag{7}$$

where:

q_{e} is the amount of MG adsorbed per unit mass of clay adsorbent (mg g^{-1}),

K_{L} is the Langmuir constant related to the adsorption capacity (L g^{-1}),

C_{e} is the concentration of MG in the solution at equilibrium (mg L^{-1}),

q_{m} is the maximum uptake dose per unit mass of clay adsorbent (mg g^{-1}),

K_{F} (mg g^{-1}) (L $mg^{-1})^n$ and $1/n$ are the Freundlich constants related to adsorption capacity and adsorption intensity, respectively,

q_{m} the Sips maximum adsorption capacity (mg g^{-1}),

K_{S} the Sips equilibrium constant (L mg^{-1}),

n the Sips model exponent describing heterogeneity.

The isotherm constants of experimental data were calculated using nonlinear regression. The isotherms constants related to Langmuir, Freundlich, and Sips models determined from the plots shown in Fig. 8 are listed in Table 3.

The results illustrated in Table 3 indicate that the parameters isotherm models are found to be suitable for fitting the present adsorption isotherms data in the following subsequent order: Langmuir > Sips > Freundlich. The Sips isotherm exponent n is near the unity indicating that the adsorption data were more of Langmuir. The monolayer adsorption capacity, q_{m}, was found to be 55.84 mg g^{-1}. The adsorption capacities of other adsorbents for MG removal from the literature are given in Table 4 for comparison. The result clearly showed that the limestone clay adsorbent is much

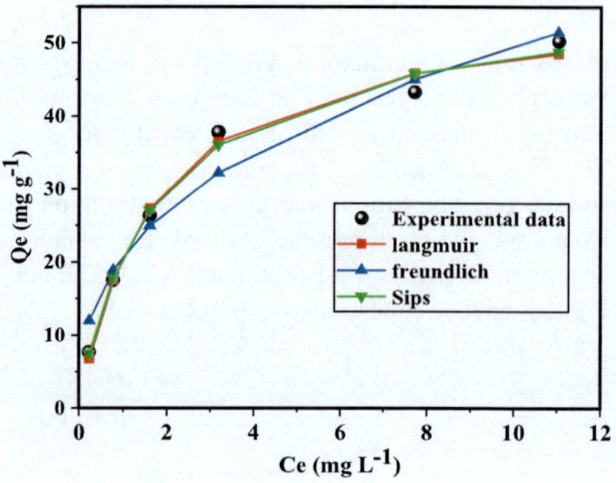

Fig. 8 Non-linearized parameter isotherm models for MG adsorption by clay

Table 3 Parameters of Langmuir, Freundlich, and Sips isotherm models for MG retention onto limestone clay

Model	Parameters	Value
Langmuir	q_m	55.84
	K_L	0.59
	R^2 (%)	99.99
Freundlich	$1/n$	0.38
	K_F	20.80
	R^2 (%)	95.57
Sips	q_m	58.47
	K_S	0.56
	n	0.92
	R^2 (%)	99.10

Table 4 Comparison of maximum adsorption capacities of various types of adsorbents for MG elimination

Adsorbants	q_m (mg g^{-1})	References
Zeolite	23.94	[4]
Zeolite	25.14	[4]
Diatomite	23.64	[18]
Diatomite	24.88	[18]
Diatomite	27.1	[18]
Cocopeat	19.2	[19]
Lignin	31.2	[7]
Humic acid	6.4	[7]
Limestone clay	55.84	This work

more effective for the removal of MG dye from an aqueous solution compared to some adsorbents previously reported.

4 Conclusions

The adsorption of MG from aqueous solutions using limestone natural clay as adsorbent has been explored in a batch system. The kinetic analysis showed that the PSO model is suitable for modeling the adsorption of MG onto clay. The adsorption isotherm study showed that the Langmuir model well fitted the experimental data with the maximum adsorption capacity found to be 55.84 mg g^{-1}. This investigation indicates that this type of clay could be a promising adsorbent for the removal of MG from aqueous solutions.

References

1. Aoulad El hadj Ali Y, Demba N'diaye A, Ahrouch M, Sakar EH, Raklami A, Lahcen AA, Stitou M (2022) Dehydrate sewage sludge as an efficient adsorbent for malachite green removal in textile wastewater: experimental and theoretical studies. Chem Afr 1–15
2. Freundlich H (1906) Over the adsorption in solution. J Phys chem 57:1100–1107
3. Guenfoud F, Mokhtari M, Akrout H (2014) Electrochemical degradation of malachite green with BDD electrodes: effect of electrochemical parameters. Diam Relat Mater 46:8–14
4. Han R, Wang Y, Sun Q, Wang L, Song J, He X, Dou C (2010) Malachite green adsorption onto natural zeolite and reuse by microwave irradiation. J Hazard Mater 175:1056–1061
5. Khattri S, Singh M (2009) Removal of malachite green from dye wastewater using neem sawdust by adsorption. J Hazard Mater 167:1089–1094
6. Langmuir I (1918) The adsorption of gases on plane surfaces of glass, mica and platinum. J Am Chem Soc 40:1361–1403
7. Lee SL, Park JH, Kim SH, Kang SW, Cho JS, Jeon JR, Lee YB, Seo DC (2019) Sorption behavior of malachite green onto pristine lignin to evaluate the possibility as a dye adsorbent by lignin. Appl Biol Chem 62:1–10
8. Mohamed A, Ghobara MM, Abdelmaksoud M, Mohamed GG (2019) A novel and highly efficient photocatalytic degradation of malachite green dye via surface modified polyacrylonitrile nanofibers/biogenic silica composite nanofibers. Sep Purif Technol 210:935–942
9. Mukherjee S (2011) Environmental mineralogy, applied mineralogy. Springer, pp 490–525
10. N'diaye AD, Ali YAEH, Bollahi MA, Stitou M, Kankou M, Fahmi D (2020) Adsorption of methylene blue from aqueous solution using Senegal River Typha australis. Mediterr J Chem 10:22–32
11. Pushpa TB, Vijayaraghavan J, Basha SS, Sekaran V, Vijayaraghavan K, Jegan J (2015) Investigation on removal of malachite green using EM based compost as adsorbent. Ecotoxicol Environ Saf 118:177–182
12. Qlihaa A, Dhimni S, Melrhaka F, Hajjaji N, Srhiri A (2016) Caractérisation physico-chimique d'une argile Marocaine. J Mater Environ Sci 7(5):1741–2175
13. Rafatullah M, Sulaiman O, Hashim R, Ahmad A (2010) Adsorption of methylene blue on low-cost adsorbents: a review. J Hazard Mater 177(1/2/3):70—80
14. Selvaraj V, Karthika TS, Mansiya C, Alagar M (2021) An over review on recently developed techniques, mechanisms and intermediate involved in the advanced azo dye degradation for industrial applications. J Mol Struct 1224:129195
15. Sinha R, Jindal R (2020) Elucidation of malachite green induced behavioural, biochemical, and histo-architectural defects in Cyprinus carpio, as piscine model. Environ Sustain Indic 8:100055
16. Srenscek-Nazzal J, Narkiewicz U, Morawski AW, Wróbel RJ, Michalkiewicz B (2015) Comparison of optimized isotherm models and error functions for carbon dioxide adsorption on activated carbon. J Chem Eng Data 60:3148–3158
17. Taghavi Fardood S, Moradnia F, Mostafaei M, Afshari Z, Faramarzi V, Ganjkhanlu S (2019) Biosynthesis of $MgFe_2O_4$ magnetic nanoparticles and its application in photo-degradation of malachite green dye and kinetic study. Nanochemistry Res 4:86–93
18. Tian L, Zhang J, Shi H, Li N, Ping Q (2016) Adsorption of malachite green by diatomite: equilibrium isotherms and kinetic studies. J Dispersion Sci Technol 37:1059–1066
19. Vijayaraghavan K, Premkumar Y, Jegan J (2016) Malachite green and crystal violet biosorption onto coco-peat: characterization and removal studies. Desalin Water Treat 57:6423–6431
20. Youssef AEHA, N'diaye AD, Fahmi D, Kankou MSA, Stitou M (2021) Adsorption of Congo red from aqueous solution using *Typha australis* leaves as a low cost adsorbent. J Environ Treat Tech 9:534–539

Ergonomic Study and Work Transformation in a Clothing Industry

Nejib Sejri

Abstract The objective of this research is to study, transform, and improve real ergonomic situations in the Tunisian clothing industry. In this context, ergonomic studies were conducted on health and safety stations (postures, repetitive gestures, musculoskeletal disorders (MSDs), their design, and the necessary manual tools). In this context, MSDs represent 70% of occupational diseases in Tunisia. Presenting the effects of these constraining factors in Tunisian clothing companies that represent more than 52% of the cases concerned. A study was conducted to identify and prove that the posture problem is related to the design of workstations. A diagnosis according to the standard of the Association for Prevention and Improvement of Working Conditions (APACT) has found that the posture problems are mainly located in the cutting rooms, through the Key Indicator Method (KIM). The Ovako Working Posture Analysis System method proves that workstation designs, as well as the repetitive gestures of the restrictive postures of many operators in clothing companies. This brings us back to the application of other ergonomic tools such as the OWAS method, as well as the standard of the German Institute for Standardization (DIN) at the level of anthropometric measurement to achieve the objectives of this study.

Keywords APACT · KIM · OWAS · MSDs · DIN · Workstation design

1 Introduction

Ergonomics is a process of analysis and transformation of work, which presents a major need by companies and their employees. Often considered as the science of "human factors", ergonomics is concerned with the prevention of the risks incurred by operators, the reliability of complex systems, the consequences of computerization and automation, and the repetition of work, on the conduct of design [1, 2]. For this scientific discipline, which aims at the fundamental understanding of the interactions

N. Sejri (✉)
Textile Engineering Laboratory of ISET Ksar Hellel, University of Monastir, Monastir, Tunisia
e-mail: sejri.nejib@gmail.com

R. Khiari and M. Jawaid (eds.), *Proceedings of the 3rd International Congress of Applied Chemistry & Environment (ICACE-3)*, Springer Proceedings in Materials 23, https://doi.org/10.1007/978-981-99-1968-0_18

of several factors and parameters that have relations with work and worker. So we must focused the importance of the study of these parameters in the field of clothing in Tunisia. In addition, occupational diseases are known as "Musculoskeletal Disorders" (**MSDs**), which in Tunisia represent a very high percentage [3, 4].

We present through this ergonomic study the lifting of loads and their movements, repetitive gestures, dimensional constraints, posture, muscular fatigue, and study of noise. Starting with the **APACT** Method for Primary Diagnosis [5, 6]: to identify different problems; using the method, **KIM** to screen for uses, handling and loads [7]; **OWAS** to know and analyze the restrictive postures and the repetitive gestures [7]; and the **DIN** standard [8] which helps us to make corrective actions at different levels (design of the positions heights) [9].

2 Methods and Materials

2.1 The APACT Method

This method is useful for diagnosis; it is well known by all the industries of the automobile it is called APACT (A: Association, P: Prevention, A: Improvement, C: Condition, T: Work). It has sheets that cover 22 ergonomics criteria in order to obtain a complete workstation profile [5, 6, 10].

The APACT sheet makes it easy to compare positions between themselves, in order to define priorities between the workstations, easy to put in place, fast, and effective (less than half a day for a complete profile). The concept of the APACT method "A systematic approach to identifying the priority problems in this diagnosis using weighting scores of from 0 to 10 or a scale simulator from 0 to 5 by a multiple of 2 (ex: −2, −1.5, −1, −0.5, 0, 0.5, 1, 1.5, 2)" for the evaluation, to eliminate the risks.

2.2 The KIM Method

It is a screening method called "Key Indicator Method", [11].

The objectives of this method according to the authors, two KIM tools have been developed for the evaluation of the following operations:

- Lift, maintain, wear;
- Push or pull a load.

This method occupies the bodily areas of the back (Table 1).

Table 1 The lifting score [11]

Lifting or moving operation (<5 s)		Hold (>5 s)		Transportation (>5 m)	
Number of operations per working day	Score	Total duration on the working day	Score	Total distance on the working day	Score
<10	1	< 5 min	1	<300 m	1
10< to 40	2	5 to 15 min	2	300 m< to 1 km	2
40< to 200	4	15 min < to 1 h	4	1< to 4 km	4
200< to 500	6	1 h < to 2 h	6	4< to 8 km	6
500< to 1000	8	2 h < to 4 h	8	8< to 16 km	8
≥1000	10	≥ 4 h	10	≥16 km	10

For lifting, holding, and moving operations, the KIM method consists primarily of evaluating a duration score from 1 to 10:

- For lifting or moving operations (<5 s), depending on the number of working days;
- For maintenance operations (>5 s), depending on the total duration on the workday;
- For displacement operations (>5 m), depending on the total distance traveled per day, scores are then evaluated according to:
  - The weight of the load: from 1 to 25, for men and for women separately;
  - The posture and the position of the load: from 1 to 8;
  - Working conditions (obstacles, space, etc.): from 0 to 2 (Fig. 1).

Table 2 shows the interpretations according to the risk classes.

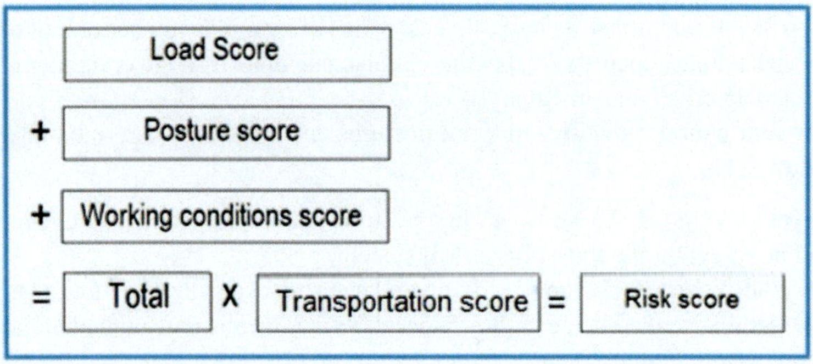

Fig. 1 Risk score calculation [11]

Table 2 Data discussion

Risk score	Risk class	Description of the work situation
<10	1	Low load, unlikely physical overload
10< to 25	2	Increased load, useful re-designing for the unfit cases[a]
25< to 50	3	Highly increased charge, recommended for re-designing
≥50	4	High load, recommended for re-designing

[a] The least fit people in this context are people aged 40 or under 21, people, new to this position and sick people

2.3 *The OWAS Method*

It is a method of evaluating the constraining postures of the human body at work; it can be used for the neck, back, upper limbs (shoulders, elbows, and hands), chest, and legs [6, 11]. It has been known since 1977 and has been improving since 1992 [12].

It is a method of analysis, semi-quantitative aims to:

- Identify and evaluate the restrictive postures at work.
- Determine the urgency of corrective actions for the position by categorizing into four categories of actions (from "no action" to "immediate corrective action").

Solutions must be made immediately to remove these postures.

This method has 3 steps:

Step 1: Video recording the workplace

Step 2: Observation analysis of video images, postures, forces, work phases; Ergonomic surveys;

The design of a new position or new method of work;

Job enhancement or work method to reduce musculoskeletal load, reduce risk and improve productivity.

Step 3: The classification into four categories of action: Compendium of observations can be done in paper form but there is also one done through computer program learning data collection and analysis.

The four action categories of work postures and work suits are to be interpreted as follows [11]:

- **Level of Action 1**: These are normal postures; do not require corrective measures and no effect on the musculoskeletal system.
- **Level of Action 2**: Postures are likely to have some specific effects on the musculoskeletal system. The constraint is weak which means that immediate action is not required. However, these situations should be corrected in the future.
- **Level of action 3**: Postures are dangerous for the musculoskeletal system. The constraints are high and the frequency of the situation should be reduced as soon as possible by acting on the working conditions.
- **Level action 4**: Postures are extremely dangerous for the system.

Table 3 shows the postures and scores of the OWAS method.

In order to identify and evaluate the inappropriate positions at work and determine the urgency of corrective measures at the station, a video recording of the work, concentrated on the selected body organs, and carried out for a representative duration.

- Video recording: This is the step of the video-filmed observation that studies the risk zone of **MSDs** at the level of each upper limb identified from an OWAS checklist. When recording you must ensure better visibility, you must take a global view, a view between the sagittal and frontal axes and make a continuous adjustment. Recording takes place at effective working times, far from breaks and without discontinuity. Therefore, it is necessary to identify a representative work period to perform this recording.
- Decomposition of work into elementary operations:

  The activity of each worker has been divided into elementary operations, while giving a number for each.
- Force level determination:

  The BORG scale is a judgmental tool of the level of subjective effort, scaled from 0 to 10 (Fig. 2).

The method consists of explaining this scale to the operators who themselves evaluate the level of force for each operation performed, relative to the body area concerned.

Table 3 The postures and scores of the OWAS method [11]

N	Upper limbs	Postures and scores
1	**For the arms**: 1. Two arms below level from the shoulders, 2. One of the arms on the level or above shoulders, 3. The two arms on the level or above **Elbows**: 3. Tended in extension, 1, 2. Upper limbs bent or in torsion **Hands and wrists**: (1, 2, 3). In extension/pressure In rotation	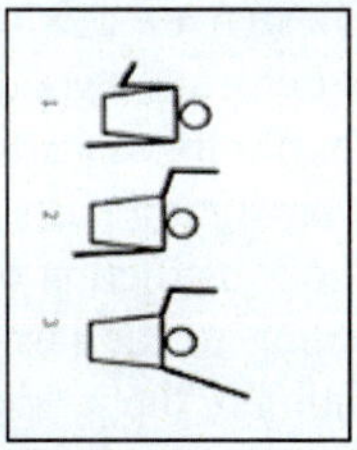 Posture upper limbs

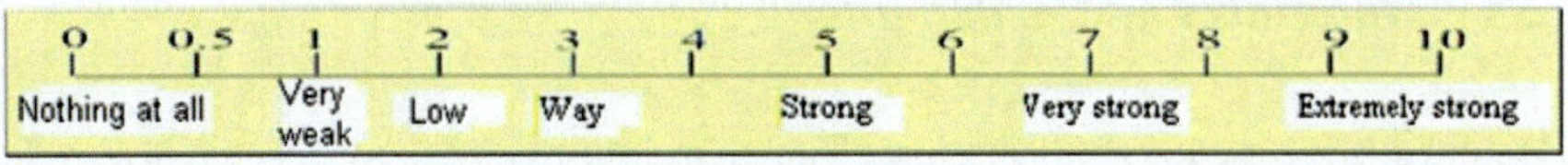

Fig. 2 BORG scale

Table 4 Body dimensions (mm)

Body dimensions (selection)		Men			Women		
		5	50	95	5	50	95
a	Size "the stature"	1650	1750	1855	1535	1625	1720
b	Eye level	1530	1630	1725	1430	1515	1605
c	Elbow relative to the support plane	1025	1100	1175	960	1020	1080
d	Forward scope up to the grip axis	685	740	815	625	690	750
e	Eye level when setting	740	795	855	705	755	805
f	Height of the elbows below the seat surface	210	240	285	185	230	275
g	Length of the lower leg, including the foot	410	450	490	375	415	450
h	Distance between the posterior and the knee	565	610	655	545	590	640

2.4 *The Design of Workstations*

The design of a workstation requires a thorough knowledge of anthropometric measurements, a good analysis of the position to be studied, as well as experience on the feasibility of tasks and a logical thinking on all the factors that have a link with the workstation posts (supply, evacuation, height, work areas, operation, handling).

The construction of the work equipment and the arrangement of posts are made according to the laws of anthropometry and physiology, as long as the mode of operation or the production process does not impose particular dimensions (Table 4 and Fig. 3) [8, 11].

2.5 *Materials and Participants*

We did this ergonomic study in a large Tunisian clothing company; located in northern Tunisia, specializing in the making of shirts. It has opened its doors to develop this work, where we have used android Sound level meters, applications Excel.

In these studies the total number of participants is 120 out of a total staff of 4 units which is about 850 people. This number is distributed as follows: 20 men and 100

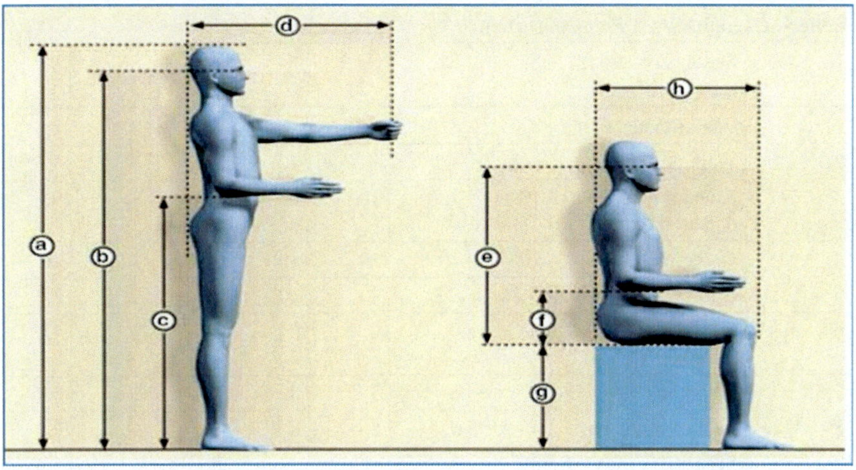

Fig. 3 Anthropometric positions [8]

women with an age range from 20 to 56 years who work in different departments and sections of the Tunisian industrial group.

3 Discussion and Results

3.1 Preliminary Diagnosis

After the diagnosis with the APACT method, and the use of the DIN standard 33 402-2 for the design of the workstations we obtained the following results; this is the APACT global score board (sections of the industrial group) which summarizes the ergonomic situation according to several parameters [13] generated by Table 5 that clarifies the critical parameters solved.

The overall score board shows that section 3 which is the cutting room that has a weight of 5.6 is a priority to deal with. In addition Table 5 shows that the work posture is poor view as it is of $1.5 < 6$ which is the minimum threshold.

3.2 Secondary Analyzes

For this analysis we used the application KIM in Excel for the screening of pull, push, and displacement of loads with a handling (Table 6).

Table 5 Overall APACT rating of sections

N	Settings/cotation sections	S1	S2	S3	S4	S5	S6	P
1	Work implantation	8	8	8	10	8	8	8.3
2	Work posture	−2	−1	−3	10	2	3	**1.5**
3	Autonomy of the operator	10	10	10	6	10	8	9.0
4	Orders	10	10	10	10	10	10	10
5	Efforts at the station	−2	−3	4	10	8	5	3.7
6	Signals and information	3	0	0	6	7	8	4.0
7	Monotony	4	6	3	8	7	3	5.2
8	Interest of the post	4	5	8	8	8	10	7.2
9	Security	8	8	−3	8	8	3	5.3
10	Work area	6	3	3	3	9	8	5.3
11	Organization of the post and doc	8	5	6	8	8	9	7.3
12	Handling	4	−3	8	8	6	8	5.2
13	Autonomy of the group	10	8	10	8	9	8	8.8
14	Dependent relationships	8	6	6	8	8	8	7.3
15	Independent relations	8	3	8	6	8	8	5.0
16	Mental charge	8	8	−3	8	3	7	5.2
17	Musical ambiance	10	8	6	10	8	10	8.7
18	Thermal atmosphere	10	10	10	10	10	10	10
19	Vibrations	10	8	6	10	6	3	7.2
20	Atmospheric	10	10	8	10	10	10	9.7
21	Brightness	8	10	8	8	6	8	8.0
22	Social environment	8	8	8	8	8	8	8.0
Weighting/section		6.8	5.7	**5.6**	8.2	7.7	7.5	6.8

S: Workshop section, P: Weighting, Minimum threshold = 6

S1: Store, S2: Padding, S3: Cutting, S4: Administration, S5: Mounting, S6: Finishing

Legend:

* 10 to 8.5: very good
* 8.5 to 6.5: good
* 6.5 to 4.5: average
* 4.5 to 2.5: inadequate
* 2.5 to 0: poor

Table 6 Description according to the risk score (KIM method)

N	Place	Pics	Problems	Recommended tools	Score from 1 to 4
1	Cutting room		Pull paper rolls with a pallet truck with a distance of 5 m and a weight of 625 kg	Using the KIM tool	**4**
2			The weight of the roll exceeds 25 kg	Using the KIM tool	

Risk scale	Risk score	Description
1	<10	Reduced load situation; improbable physical overload
2	10 to <25	Increased load situations, possible physical overload in people who are less resistant to this group, a redevelopment of the workstation is useful
3	25 < 50	Highly increased load situation, physical overload also possible in normal people. For this group, a rearrangement of the workstation is recommended
4	> to 50	High load situation, probable physical overload. A rearrangement of the workstation is necessary

We can say that we have in this case a very serious risk situation because it is enough to make a false movement to have a muscular lesion or a breakdown in the muscles. A rearrangement of the workstation is necessary.

3.3 *Noise Analysis*

It is a set of sounds produced by vibrations more or less irregular, any phenomenon perceptible by hearing. In this project we used the European standard NF EN ISO 9612 [14], which is adopted by several countries, to study noise. It also provides a step-by-step approach for determining the level of noise exposure in the workplace by measuring noise levels. The method includes the following main steps: Work analysis, selection of a measurement strategy, error handling and uncertainty assessment, calculations, and presentation of the results.

In our work we used the android application (sound level meter) calibrated against the professional sound level meter of the technical department of occupational medicine. This stady shows that the sound level of a worker who performs several stains such as embroidery machine, lids preparation, and control of embroidered pieces is important. We measured the sound of a single head and found 70 dB (A) according to the standard.

We used the formula of the NF ISO 9612 standard and we found the following results: The total sound level noise in dB = 70 dB + 10 * log(19) = 70 + 10 * 1.27 = 82.78 dB ≈ 83 dB.

According to this study, we can say that the embroidery machine is high noise level, so we must think of solutions to minimize its noise level.

After these analyzes we have proposed some solutions that can help operators in their work is more comfortable, so to have more comfort at the postures, gestures, and tools (Table 7).

Table 7 Minimize noise by separating aluminum

Before	After
	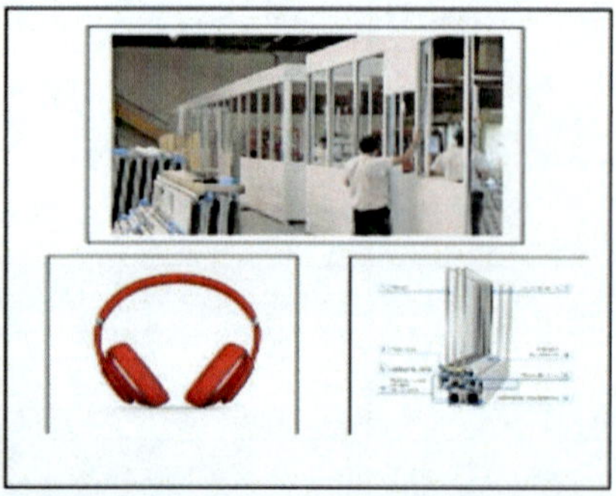
Noise provided by this machine = 83.4 dB (A) which causes the Suffering of workers who work near this machine	Isolate this machine from the workshop by designing an aluminum block with profiled 3 insulating layers giving the equipment of the individual protections for the workers
Profit	
To minimize the tiredness, the headaches of the workers who do the manual sewing due to this machine, to improve the productivity and the noise level will be about 70 dB (A)	

4 Conclusions

In this paper, the methods and ergonomic tools are presented; APACT, KIM, OWAS, and DIN, the EN ISO 9612 standard to solve posture problems, repetitive gestures, workstation design, lighting, noise, health, and safety within Tunisian apparel companies.

The participation and commitment of management, using the participatory approach, gives importance to this work, as well as the recommended solutions that will be an added value which is reflected by the improvement of more than 50% at the level of illumination and gain of energy to 10%, reduction of the risk of MSDs to 21%. In addition, we can have a reduction of noise of 15 dB through the solution of the box of Aluminum with window glass anti-noise.

References

1. Marie B, et al (2002) Etudes et recherches. In: Institut de recherche Robert-Sauvé en santé et en sécurité du travail, Montréal Québec, Canada
2. Lakhal A, Sejri N, Jaafar F, Chaabouni Y, Cheikhrouhou M (2019) Elaboration of a rest index for a sewing post in a clothing industry. Ind Text 70(5):430–434
3. Ghram R, Fournier C, Khalfallah T, Francis S (2010) Analysis of socio-cultural factors and occurrence of disorders musculoskeletal: the case of dressmakers in Tunisia. journals.openedition.org/pistes. https://doi.org/10.4000/pistes.2459
4. NHIF (2015) National Health Insurance Fund, Tunisia
5. Rose JP et al (2017) The experimental method in psychology. In: Int J Clin Health Psychol 17(2):120–127
6. APACT (1991) Method study of the association for the prevention and improvement of working conditions AFPI, 56 street of Wagram—75854 PARIS cedex 17, pp. 35–37
7. Malchaires J (2005) Notes malchaires risk, France, pp 30–33
8. Jürgens WH (2004) DIN 33 402—part 2 Dortmund/Berlin/Dresden, Germany
9. Kamusella C (2015) Anthropometric measures. In: Deutsch Institute standard (DIN 33 402-2)
10. Alfredo L (2011) Ergonomic Steelcase Euro Institute Strasbourg Area, France
11. Malchaires J (2011) A classification of methods for assessing and/or preventing the risks of musculoskeletal disorders. European Trade Union Institute. ISBN 978-287452-223-9
12. Malchaires J, Gauthy JR, (2011) Guide classification of evaluation methods and/or risk prevention musculoskeletal disorders. European Trade Union Institute. ISBN 978-2-87452-223-9
13. Maset C (2012) Health and safety manager, Health and safety committee of Gesim Ergonomics in business. United States Department of Labor, Metz Area, France
14. AFNOR (2009) NF ISO 9612

Physicochemical Compositions of Date Palm (*Phoenix Dactylifera* L.) Seeds Growing in Mauritania: Preliminary Study

Sid'Ahmed Fall Jiddeine, Mohamed Moutaly, Ould El Moustapha Abdallahi, Sidi Mohamed Laghdaf, Mohamed Sid'Ahmed Kankou, and Abdoulaye Demba N'diaye

Abstract In this study, the proximate and mineral compositions of the seeds of three varieties (El hemer, Tiguidert, and El kerkar) of date palm (*Phoenix dactylifera* L.), growing in Mauritania, were analyzed. This work showed that the moisture content varies between 0.26 and 0.30%, the ash content ranges from 1.26 to 1.45%, the protein content is between 9.6 and 17.5%, and the fat content is between 8.66 and 10.46%. Among the six studied minerals, magnesium, calcium, and potassium were the predominant macroelements and iron was found to be the predominant microelement. The obtained results showed that date seeds could be used as ingredients to improve the nutritional value of some functional foods for human consumption as well as using additives in food, pharmaceutical, and cosmetic industries.

Keywords Physicochemical · Proximate · Nutritional · Date seeds · Mauritania

1 Introduction

The date palm is a monocotyledonous tree, of the family Palmaceae or Phoenicicaceae, known as *Phœnix dactylifera* L. [10]. The word Phoenix is derived from the name of the date tree by the ancient Greeks who considered it to be the tree of the

S. F. Jiddeine · M. S. Kankou · A. D. N'diaye (✉)
UR Eau Pollution et Environnement, Département de Chimie, Faculté des Sciences et Techniques, Université de Nouakchott, BP 880 Nouakchott, Mauritanie
e-mail: abdouldemba@yahoo.fr

M. Moutaly · S. M. Laghdaf
Laboratoire de Chimie, Institut National de Recherches en Santé Publique, BP 695 Nouakchott, Mauritanie

O. E. M. Abdallahi
Laboratoire de Biotechnologie Alimentaire, Département des Sciences et Technologies des Aliments, Institut Supérieur d'Enseignement Technologique, Rosso, Mauritanie

R. Khiari and M. Jawaid (eds.), *Proceedings of the 3rd International Congress of Applied Chemistry & Environment (ICACE–3)*, Springer Proceedings in Materials 23, https://doi.org/10.1007/978-981-99-1968-0_19

Phoenicians (from the Greek "Phoen" blood red, characteristic of their skin color) or Phoenicians. While "Dactylifera" comes from the Latin "dactylus" originating from the Greek "daktylos" meaning finger by resemblance with the shape of the date fruit [6]. The date palm tree is qualified by several names depending on the language considered, thus it is named: date palm in French, Nakhla in Arabic, Palmadatilera (Spanish), and Palma dattero (Italian) [21].

The date palm tree can adapt to variable climatological conditions. Naturally, it is thermophilic; requiring a warm, dry, and sunny climate. It also adapts to all soils. However, it is very sensitive to moisture during the pollination period and during ripening [1]. Palms are some of the most useful plants in the agricultural economy of the countries of the phoenicultural zones. It has always played an important economic and social role for the populations of these regions.

The cultivation of date palm is mainly located in the arid regions south of the Mediterranean and in the southern part of the Middle East from the South of Iran in the East to the Atlantic coast of North Africa in the West, between the latitudes 35° North and 15° South. Spain is the only country in Europe that produces dates mainly in the famous palm grove of Elche, located west of Alicante at 39° North. The date palm is also cultivated on a smaller scale in Mexico, Argentina, Australia, and the United States of America.

Mauritania is one of the date-producing countries in the world. The palm grove is composed of about 1.87 million palm trees spread over the 218 oases listed. These oases, which cover an estimated area of about 12,000 ha, are distributed in the regions of Adrar, Tagant, Assaba, and the two Hodhs with a density of 171, 354, 504, and 150 palm trees per hectare, respectively [16].

The date is the fruit of the date palm. It is a berry, generally elongated, its dimensions are very variable from 1.5 to 8 cm in length and a weight of 2–20 g. Its color goes from the yellowish white to the dark very dark almost black, passing by amber, red, and brown. The date contains a single seed called "stone". The edible part of the date is called "flesh" or "pulp".

The use of the main product of the sector, dates, produces two types of by-products: non-consumable dates and date seeds. Date seeds can be discarded or used in animal feed. However, some studies have shown that date seeds contain a wide range of nutritional and therapeutic functional compounds [2, 4, 11, 13, 17, 20].

The purpose of this preliminary study was to evaluate the proximate and mineral composition of the seeds of three varieties of Mauritanian date palm, includingthe content of protein, moisture, ash, fat, and minerals such as calcium, magnesium, potassium, sodium, iron, and zinc.

2 Materials and Methods

2.1 Sample Collection and Preparation

The dates were obtained from Tevragh Zina market, Nouakchott City in Mauritania. These three varieties of dates come from two different areas: the El Hemer variety originates from Aoujeft and the other two varieties, that is to say, El Kerkar and Tiguidert come from Toungad. Seeds were separated directly from the fruits of each date variety (Tiguidert, Elhemmer, and Elkerkar). The seeds were subsequently washed, dried, and ground into a fine powder. This powder was then stored in a desiccator before use.

2.2 Analytical Procedures

Physical properties of date seeds: From each variety, twenty date fruits were taken at random to determine the date weight and seed length, width, and weight.

Physicochemical properties: The technique described by Karaye et al. [15] was utilized to evaluate the moisture. The method outlined by AOAC [5] was applied to evaluate the proteins and ash. Lipids were calculated in accordance with the method outlined by Folch et al. [12]. Mineral ions like calcium (Ca), magnesium (Mg), sodium (Na), potassium (K), iron (Fe), and zinc (Zn) were identified using this method: 0.5 g of the dates seed powder was placed in a Teflon digestion tube. 7 mL of 65% concentrated HNO_3 and 1 mL of 30% H_2O_2 were also added. The tubes are then introduced into the microwave mineralizer (Aton Paar, France) until complete mineralization. The mineralization solutions are transferred into 50 mL flasks and then filled up with distilled water. These solutions were used for the determination of mineral content. Calcium, magnesium, zinc, and iron were determined by Atomic Absorption Spectrophotometer (AAS PGG 990), while potassium and sodium were determined by Flame Photometric FP 640.

3 Results and Discussion

3.1 Physical Properties of Date Seeds

Table 1 presents the results of the physical characterization of seeds of different varieties. The results presented clearly show significant differences in weight. Indeed, for the weight of the date palm seeds, a variation between a value of 0.803 g for the Tiguidert variety and a value of 1.112 g for the Elhemer variety was noted. Concerning the results of the length of the date seeds a variation between a length of 1.840 cm

Table 1 Date seeds physical properties

	EL Hemer	Tiguidert	EL Kerkar	Mean
Weight date (g)	5.948	4.520	3.070	4.51 ± 1.44
Weight seed (g)	1.112	0.803	0.991	0.97 ± 0.16
Length of seed (cm)	2.230	1.840	1.995	2.02 ± 0.19
Diameter of seed (cm)	0.685	0.580	0.650	0.64 ± 0.05

for the seeds of the Tiguidert variety and a length of 2.230 cm for the seeds of the Elhemer variety was observed. In addition, the highest diameter of date seeds was found to be 0.685 cm and the lowest was 0.580 cm for seeds of the Tigaidart variety. These results are slightly in line with those published by Al Juhaimi et al. [3]. This characteristic is extensively utilized by farmers to assess the quality of varieties. In addition to the effect of cultivars, the physical properties mentioned above can be influenced by climatic conditions and some cultivation practices [9].

3.2 Physicochemical Properties of Date Seeds

The respective ash content of Elhemer, Tiguidert, and Elhemer seeds are 1.45%, 1.26%, and 1.32%, respectively (Table 2). The protein content of 9.6% and 17.5% was found in the Elhemer and Elkerkar seeds, respectively. The highest contents of protein (17.5%) and fat (10.46%) were shown in Elkerkar seeds. The lowest contents of fats (8.66%) were found in the case of seeds of the Tiguidert variety. This variety contains the highest moisture content with a value of 0.30%. The lowest rate of proteins, which is equal to 9.6%, was observed in the case of seeds of the Elhemer variety that contained the lowest moisture content, which is equal to 0.26%. The lowest yield of proteins (9.6%) was found in Elhemer variety seeds with the highest ash content (1.45%).

Table 3 shows the comparison of the fat contents of the seeds of the present study with those found in the literature. From Table 3, we can say that the fat contents of the different seeds of the three different varieties of date of the present study are superior or comparable to those cited in the literature. These very high fat levels recorded in our study indicate that the seeds of the three Mauritanian varieties could be a good source of edible vegetable oil.

Table 2 Proximate compositions content of Mauritanian date seeds

	El Hemer	Tiguidert	El Kerkar
Moisture (%)	0.26	0.30	0.28
Ash (%)	1.45	1.26	1.32
Proteins (%)	9.6	15.8	17.5
Fat (%)	9.33	8.66	10.46

Table 3 Comparison of the fat content obtained by the seeds of the three varieties of date with those of other varieties of date cited in the literature

Varieties	Fat (%)	References
Khalas	10.5	[2]
Lulu	10.5	[14]
Deglet Nour	10.1	[8]
Fardh	8.2	[18]
Sokkery	6.5	[18]
El Hemer	9.33	This study
Tiguidert	8.66	This study
El Kerkar	10.46	This study

Table 4 Comparison of the protein content obtained by the seeds of the three varieties of date seeds with those of the other varieties of seeds

Varieties	Proteins (%)	References
Khalas	5.7	[2]
Sokkery	6.4	[18]
Shabebe	4.8	[18]
Maktoomi	5.8	[18]
Raziz	6.9	[18]
El Hemer	9.6	This study
Tiguidert	15.8	This study
El Kerkar	17.5	This study

Table 4 shows the comparison of the protein contents of the seeds of the present study with those found in the literature. According to Table 4, these values are higher than those indicated in the literature. These very high protein levels recorded in our study indicate that the seeds of the three Mauritanian varieties could be an important additional source of protein.

The levels of six minerals, namely Ca, Mg, K, Na, Fe, and Zn, which are usually analyzed, were assessed for the three varieties (El hemer, Tiguidert, and Elkerkar) of date seeds. From the results presented in Table 5, the following observations can be derived:

Table 5 Minerals composition of date seeds (mg/100)

Mineral elements	El Hemer	Tiguidert	El Kerkar
Ca	29	24	46
Mg	35	35	34
K	180	170	170
Na	21	22	22
Fe	36	25	28
Zn	26	26	29

- Potassium was the highest mineral; it varied between 170 mg/g in the case of the Tiguidert and the Elkerkar varieties and 180 mg/g in the case of the Elhemer variety seeds, followed by calcium, which varied between 24 and 46 mg/g and magnesium which varied from 34 to 35 mg/g.
- The lowest level was observed for the mineral sodium, which ranged from 21 mg/g for Elhemer variety seeds to 22 mg/g for Tiguidert and Elkerkar varieties seeds.
- With regard to microelements, iron was present at a high level with a value ranging from 36 mg/g of the Elhemer variety seeds to 25 mg/g for seeds of the Tiguidert variety. In addition, zinc, with results fluctuating between 26 and 29 mg/g, was the lowest microelement.

These observations show a clear variation in the mineral content of the different date seeds of the studied varieties. This variation may be due to one or more of the following factors: variety, soil type, climatic conditions, irrigation water, and fertilizers [3, 13, 19].

A study by Batra and Seth [7] revealed that micronutrient deficiencies represent a major public health problem in many developing countries, especially for infants and pregnant women because they need sufficient amounts of micronutrients to maintain normal growth and development. The high content of micronutrients present in the seeds of dates can encourage the food industries to use date palm seeds to fortify their products.

4 Conclusions

This preliminary research analyzed the physicochemical composition of the seed of date. The results of the proximate and minerals analysis showed that the seeds from date fruits could be considered rich in minerals elements. Due to its high mineral content and excellent ready-to-use composition, this by-product is widely used as an ingredient to improve the nutritional value of certain functional foods and is used as a medicinal oil for human consumption.

References

1. Abu-Rida IM, Gil-Izquierdo Á, Medina S, Ferreres F (2017) Phenolic composition profiling of different edible parts and by-products of date palm (*Phoenix dactylifera* L.) by using HPLC-DAD-ESI/MSn. Food Res Int 100:494–500
2. Al-Farsi M, Alasalvar C, Al-Abid M, Al-Shoaily K, Al-Amry M, Al-Rawahy F (2007) Compositional and functional characteristics of dates, syrups, and their by-products. Food Chem 104:943–947
3. Al Juhaimi F, Ghafoor K, Ozcan MM (2012) Physical and chemical properties, antioxidant activity, total phenol and mineral profile of seeds of seven different date fruit (*Phoenix dactylifera* L.) varieties. Int J Food Sci Nutr 63:84–89

4. Al Qarawi AA, Abdel-Rahman H, Ali BH, Mousa HM, ElMougy SA (2005) The ameliorative effect of dates (*Phoenix dactylifera* L.) on ethanol-induced gastric ulcer in rats. J Ethnopharmacol 98:313–317
5. AOAC (1995) Official methods of analysis, 16th edn. Association of Analytical Chemists, Washington, DC. Afr J Food Sci 4:115–119
6. Arroussi A, Tahri A, Kalloum S, Abdelli N, Bouziane N (2022) Valorization of lignocellulosic material from date palm waste (*Phoenix dactylifera* L.) Elhmira cultivar by composting as organic fertilizer. UPB Sci Bull, Ser B 84(1):59–72
7. Batra J, Seth PK (2002) Effect of iron deficiency on developing rat brain. Indian J Clin Biochem 17(2):108–114
8. Besbes S, Blecker C, Deroanne C, Drira NE, Attia H (2004) Date seeds: chemical composition and characteristic profiles of the lipid fraction. Food Chem 577–584
9. Bouhlali ET, Alem C, Ennassir J, Benlyas M, Mbark AN, Zegzouti YF (2017). Phytochemical compositions and antioxidant capacity of three date (*Phoenix dactylifera* L.) seeds varieties grown in the South East Morocco. J Saudi Soc Agric Sci 16:350–357
10. Chehma A, Longo HF (2001) Valorisation des Sous-Produits du Palmier Dattier en Vue de leur Utilisation en Alimentation du Bétail. Rev Energ Ren: Production et Valorisation—Biomasse 59–64
11. Doha MA, Al-Okbi SY (2004) In vivo evaluation of antioxidant and anti-inflammatory activity of different extracts of date fruits in adjuvant arthritis. Pol J Food Nutr Sci 13:397–402
12. Folch J, Lees M, Sloane Stanley GH (1957) Une méthode simple pour l'isolement et purification des lipides totaux des tissus animaux. J Bouillir Chem 226(1):497–509
13. Habib HM, Ibrahim WH (2009) Nutritional quality evaluation of eighteen date pit varieties. Int J Food Sci Nutr 60:99–111
14. Hassan LG, Sokoto MA, Dangoggo SM, Ladan MJ (2006) Proximate, amino acid and minerals composition of silk cotton seeds (Ceiba pentandra) Afr J Sci 9:29–35
15. Karaye IU, Ladan MU, Adili IS, Shehu A, Lawal HM, Sahabi MH (2020) Phytochemistry and proximate composition of fruit pulp and seed of desert date, Balanites aegyptiaca (Del.). Int J Sci Glob Sustain 6(2)
16. Kneyta M, Doulbeau S (2012) Le palmier dattier en Mauritanie. IRD Editions, Open Edition Book License
17. Mrabet A, Ferchichi A, Chaira N, Mohamed BS, Baaziz M, Penny TM (2008) Physico-chemical characteristics and total quality of date palm varieties grown in the Southern of Tunisia. Pak J Biol Sci 11:1003–1008
18. Quinsac A, Bazin De Caix C-P, Bernard C, Brosson B, Carre P, Chardon G, Simonneau D (2013) Comparaison des méthodes de détermination de la matière grasse dans les tourteaux d'oléagineux—Étude collaborative inter-laboratoires. OCL 20(5):503
19. Rahman MS, Kasapis S, Al-Kharusi NSZ, Al-Marhubi IM, Khan AJ (2007) Composition characterization and thermal transition of date pits powders. J Food Eng 80:1–10
20. Salah S, Al-Maiman SA (2005) Effect of date palm (*Phoenix dactylifera*) seed fibers on plasma lipids in rats. J King Saud Univ 17:117–123
21. Vico A, Pérez-Murcia MD, Bustamante MA, Agulló E, Marhuenda-Egea FC, Sáez JA (2018) Valorization of date palm (*Phoenix dactylifera* L.) pruning biomass by co-composting with urban and agri-food sludge. J Environ Manag 226:408–415